# Residual Shear Behavior of Composite Soils

# Residual Shear Behavior of Composite Soils

Yanrong Li
*Taiyuan University of Technology, Taiyuan, China*

CRC Press
Taylor & Francis Group
Boca Raton London New York

CRC Press is an imprint of the
Taylor & Francis Group, an **informa** business
A BALKEMA BOOK

CRC Press
Taylor & Francis Group
6000 Broken Sound Parkway NW, Suite 300
Boca Raton, FL 33487-2742

First issued in paperback 2019

© 2017 by Taylor & Francis Group, LLC
CRC Press is an imprint of Taylor & Francis Group, an Informa business

Typeset by MPS Limited, Chennai, India

No claim to original U.S. Government works

ISBN-13: 978-1-138-03287-3 (hbk)
ISBN-13: 978-0-367-88568-7 (pbk)

This book contains information obtained from authentic and highly regarded sources. Reasonable efforts have been made to publish reliable data and information, but the author and publisher cannot assume responsibility for the validity of all materials or the consequences of their use. The authors and publishers have attempted to trace the copyright holders of all material reproduced in this publication and apologize to copyright holders if permission to publish in this form has not been obtained. If any copyright material has not been acknowledged please write and let us know so we may rectify in any future reprint.

Except as permitted under U.S. Copyright Law, no part of this book may be reprinted, reproduced, transmitted, or utilized in any form by any electronic, mechanical, or other means, now known or hereafter invented, including photocopying, microfilming, and recording, or in any information storage or retrieval system, without written permission from the publishers.

For permission to photocopy or use material electronically from this work, please access www.copyright.com (http://www.copyright.com/) or contact the Copyright Clearance Center, Inc. (CCC), 222 Rosewood Drive, Danvers, MA 01923, 978-750-8400. CCC is a not-for-profit organization that Provides licenses and registration for a variety of users. For organizations that have been granted a photocopy license by the CCC, a separate system of payment has been arranged.

**Trademark Notice:** Product or corporate names may be trademarks or registered trademarks, and are used only for identification and explanation without intent to infringe.

*Library of Congress Cataloging-in-Publication Data*

**Visit the Taylor & Francis Web site at**
**http://www.taylorandfrancis.com**

**and the CRC Press Web site at**
**http://www.crcpress.com**

# Table of contents

# Foreword

I am pleased to write a foreword for this unique reference on the residual shear strength of artificial soil mixtures and the composite soils of three large landslides in the Three Gorges Reservoir area. The book starts with an extensive and in-depth coverage of published literature on the relevant aspects of soils' shear behavior and strength. The experimental methods of determining residual shear strength of granular and cohesive soils are summarized together with the supporting classification tests and particle shape quantification methods. The results of tests on artificial soils and the specific test conditions adopted are presented and systematically discussed to construct a conceptual framework to discuss the behavior of natural composite soils from landslide slip zones. This leads to a number of important observations and concrete conclusions about the factors influencing the shear behavior. The results of tests on the slip zone soils are then presented and analyzed with reference to those of the controlled soil mixtures. This study also represents one of the largest experimental campaigns in terms of the ranges of soil and test conditions under which tests were carried out. One of the most novel aspects of this study is the introduction of signal analysis methodology to investigate the fluctuations during the residual shear of granular and composite soils. These aspects make this book an indispensible reference in graduate teaching and research and particularly in the practice of stabilizing dormant landslides.

The research results presented in this book are partially based on a project carried out by the author initially under my supervision during my tenure at the University of Hong Kong. The author, Professor Li, has continued to work on the topic in past years and published most components of his research in a variety of journals. This book successfully combines these research outputs making it a one-stop reference on residual shear strength of granular and composite soils.

Adnan Aydin PhD, PE
Professor of Geology and Geological Engineering
University MS, USA

# Preface

In the area of the Three Gorges Project (TGP), more than two thousand ancient landslide masses of various scales have been identified in earlier field investigations. The failure potential of these landslides poses a tremendous threat to navigation in the reservoir and normal operation of the TGP dam. For mitigating the landslide disasters in this area, much attention should be devoted to the study of the fundamental mechanical properties, especially the residual shear behavior, of these slip zones. However, unlike cohesive soils which have been widely investigated for decades, slip zone soils in these landslides have not been well studied and documented due to their distinct properties, in particular: large range of particle size; abundant coarse fraction; and diverse properties of coarse particles. This type of soil is called composite soil in this book.

This book has the purpose of developing an understanding of the factors determining and influencing the shear behavior of soils, with emphasis on composite soils, as they are the most encountered materials in geological and geotechnical engineering in mountainous areas. This objective is reached by examining the soil compressibility, structure of shear zone and its evolution, and water content of shear zone and shear mode of soils together with analyses of the influences of intrinsic properties, e.g. Atterberg limits, particle size distribution, particle shape, and testing conditions, e.g. normal stress and shearing rate.

An in-depth review is presented in an approximately chronological order and covers almost all the factors that are believed to influence the mechanical behavior of soils. The equipment and test techniques for shear strength of soils are detailed. The residual shear behavior of composite soil is investigated by means of a systematic laboratory testing program using a large ring shear apparatus and an intermediate direct shear box. The Fast Fourier Transform is employed for the first time to analyze the fluctuations of measured shear stress and discovers the close relationships with both intrinsic properties of soils and testing conditions.

Although the book is aimed primarily at researchers in geological and geotechnical engineering, it contains material of interest to students of geology and soil science and also should be a useful reference for practicing engineers faced with composite soils.

Yanrong Li
Taiyuan University of Technology, Taiyuan, China
July 2016

# Acknowledgements

The journey for pursuing this book would not have been possible without the perpetual support and encouragement of Professor Adnan Aydin from the University of Mississippi, Professor Lung Sang Chan and Dr. Albert T. Yeung from the University of Hong Kong. I would like to express the utmost gratitude to them for their invaluable suggestions, critical and stimulating discussions on the subject.

I cannot thank enough my family for their encouragement whenever I was frustrated. My special thanks are owed to my wife Ms. Xiaohong Deng for her understanding, concern, inspiration and loving support. The sweetest experiences that we had in this period were the births of our sons Haoqian and Haoze. They bring a new perspective to our life mission.

Special gratitude is given to Professor Baoping Wen for her help and suggestions in the early stage of the research. Many thanks are also expressed to my colleagues and students for their assistance with some of the figures. I would like to thank the Head of the Geotechnical Engineering Office and the Director of the Civil Engineering and Development Department, the Government of the Hong Kong Special Administrative Region, for the permission to reproduce figures in their publication. The permission for figure reproduction from Chinese Journal of Rock Mechanics and Engineering and Chinese Journal of Engineering Geology are also greatly appreciated.

# About the author

Prof. **Yanrong Li** works in the Department of Earth Sciences and Engineering at Taiyuan University of Technology, China. He is the founding Director of the Collaborative Innovation Center for Geohazard Process and Prevention at this university. He worked for about ten years as a practicing Engineer and field Geologist in Hong Kong, Australia and China before he turned to be a full Professor. The research of Prof. Yanrong Li focuses on soil and rock mechanics, slope and tunnel engineering, geohazard investigation and mitigation, and geomorphology. He is a Chartered Geologist (CGeol) of Geological Society of London and a Professional Member (MIMMM) of the Institute of Materials, Minerals and Mining. He served as an Associate Editor of KSCE – Journal of Civil Engineering and is now an Editorial Board Member of Bulletin of Engineering Geology and the Environment. Prof. Yanrong Li (li.dennis@hotmail.com) welcomes discussions on the contents of this book and diverse collaboration with experts of same interest.

# List of tables

# List of figures

# List of Symbols

The symbols often used in the book are defined below, while those symbols seldom used are defined where they appear in the text for the first time.

$A$ = wave amplitude
$c'_R$ = effective cohesion intercept
$c'\mu$ = undrained cohesion of fine matrix
$CF$, $F_{clay}$ = clay fraction (% <2 μm)
$C$ = volumetric grain concentration ratio
$C_c$ = coefficient of curvature
$D_{50}$ = particle size for which 50 percent of particles are finer
$D_{10}$ = particle size for which 10 percent of particles are finer
$e$ = void ratio
$e_g$ = granular void ratio
$e_{min}$ = void ratio at the minimum density
$e_{max}$ = void ratio at the maximum density
$E_k$ = kinetic energy
$E_w$ = wave energy
$FG$ = mixture of kaolin, silt and coarse fraction
$FN$ = mixture of kaoline, silt, river sand and crushed gravels
$G_s$ = specific gravity
$I_B$ = brittleness index
$I_P$ = plasticity index
$KG$ = mixture of kaolin and glass materials
$M_g$ = mass of gravel
$M_s$ = mass of sand
$M_m$ = mass of silt
$p'$ = mean effective stress
$q$ = deviatoric stress
$r_{vc}$ = volume ratio (ratio of volume of fine matrix to the total volume of the soil)
$R$ = residual factor
$R_g$ = specific gravity of gravel
$R_s$ = specific gravity of sand
$R_m$ = specific gravity of silt
$\tan\varphi'$ = effective friction coefficient
$\tan\varphi_R$ = residual friction coefficient

$\tan \varphi_P$ = peak friction coefficient
$w$ = water content
$\theta_{max}$ = maximum angle of dilation
$\lambda_R$ = brittleness parameter
$\mu$ = pore water pressure
$\sigma$ = total normal stress
$\sigma'$, $\sigma'_n$ = effective normal stress
$\sigma'_p$ = preconsolidated pressure
$\tau$ = shear stress
$\tau/\sigma'$ = stress ratio
$\varphi$ = total friction angle
$\varphi'$ = effective friction angle
$\varphi_c$ = residual friction angle of clay
$\varphi_{c-s}$ = residual friction angle of clay-sand mixture
$\varphi_s$ = residual friction angle of sand
$\varphi'_P$ = peak effective friction angle
$\varphi'_R$ = drained residual friction angle
$\varphi_u$ = interparticle friction angle
$[\phi'_i]s$ = secant friction angle
$[\phi'_R]s$ = secant fully softened friction angle
$\omega$ = angular frequency
$\omega_L$ = liquid limit
$\omega_P$ = plastic limit

# Chapter 1

# Introduction

## 1.1 BACKGROUND

Landslides are the most common type of geohazards in mountainous areas. They have tremendous socio-economic impact, especially in developing countries (Brabb & Harrod, 1989; Wang, 1992; Schuster, 1994). In a broad sense, landslide is a general term applied to all modes of downslope movement of a mass of soil and rock material, such as slides, debris flows, rockfalls, topples, etc. (Varnes, 1978; Hutchinson, 1988; Cruden & Varnes, 1996). In a narrow sense, landslide usually refers to the slide mode of soil and/or rock mass movement, in which there is a distinct surface or zone at the contact between the moving mass and the underlying stable materials (Zaruba & Mencl, 1982; Wen & Aydin, 2003; Cornforth, 2005). In this book, the term landslide is used only in this narrow sense.

China, with its mountainous terrain and high population density, is one of the countries, which suffers most from landslides (Leroueil, 2001). Li and Wang (1992) conservatively estimated that during the period from 1951 to 1989, landslides have caused more than 5000 deaths and about US$20 billion in economic losses in China. In the TGP (Three Gorges Project) area, approximately 2,490 ancient/old landslides of various scales and 90 gullies created by mud-rock flows have been identified, and more than 70 landslide failures occurred since 1982, causing at least 363 fatalities (Li, 2002; Dai *et al.*, 2004; Liu *et al.*, 2004; Wen *et al.*, 2007). With the reservoir water level rising up to its maximum operational level of 175 m after the completion of the dam construction, some of these existing landslide masses were reactivated and new landslides were triggered. A vivid example of this is the Qiangjianping landslide, which occurred just 40 days after the first phase of the impoundment raised the water level from 95 m to 135 m, though no pre-existing landslide trace on this slope had been identified during the earlier field investigations in the TGP area (Dai *et al.*, 2004; Wang *et al.*, 2004).

It is well known that slip zones develop along the weakest parts in soil masses. Slip zones of ancient/old landslides often appear to have undergone stronger physical and chemical deterioration than the host soils, making their mechanical properties special (Cornforth, 2005). Although it has long been recognized that investigations of slip zones are of great importance for better understanding and mitigating landslide disasters, this goal has not yet been achieved fully due to the following three reasons (Hutchinson, 1988; Cruden & Varnes, 1996; Dikau *et al.*, 1996; Wen & Aydin, 2003).

Soils forming slip zones significantly differ from common soils, such as alluvium, colluviums and saprolite, etc., in terms of its heterogeneity, origin and structure. The diverse nature of slip zone soils largely originates from geological aspects of the host soil or rock mass, including mineralogy, lithology, structural geology, hydrogeology and active superficial geological process (weathering, deposition, and erosion, etc.).

It is extremely difficult to locate and to sample slip zones during site investigations. In practice, slip zones usually cannot be determined or detected with certainty, even when the site investigation has sufficient budget and advanced technology. This difficulty increases with the landslide volume and the complexity of original geomorphology. In addition, sampling of slip zones to obtain undisturbed samples for sophisticated laboratory tests also presents practical difficulties, even when there are exploratory adits/tunnels or test pits in which slip zones may partly be exposed.

Conducting laboratory and in-situ tests on slip zone soils is usually time-consuming and costly. The adverse location of slip zones makes it difficult to conduct in-situ tests, and it commonly takes several days before samples are brought to the laboratory for sequential tests.

Among all properties of slip zone soils, their residual strength has been recognized as the most important mechanical property associated with the stability and reactivation of ancient landslides (Skempton, 1964; Lupini, 1980). Since the 1960s, both drained and undrained residual strength were extensively studied in the laboratory, using a variety of experimental techniques, with the assumption that large displacements lead to the residual condition (Skempton, 1964; Kenney, 1967, 1977; La Gatta, 1970; Garga, 1970; Bishop *et al.*, 1971; Vaughan *et al.*, 1976; Bucher, 1975; Lupini *et al.*, 1981; Lemos, 1986; Tika, 1989). Bishop (1971) reported the development of a small ring shear apparatus for measuring residual strength of cohesive soils, which allows displacements large enough to reach residual conditions. Besides this, torsional shear, hollow cylinder, reversal direct shear and ring shear apparatuses have also commonly been used to examine the residual strength. La Gatta (1970) and Lupini *et al.* (1981) presented extensive reviews of the facilities used for this purpose. By the aid of these apparatuses, cohesive soils were widely investigated and profound results and findings were achieved. Many influential factors were recognized, such as Atterberg limits, particle size distribution and clay mineralogy, etc., and relationships between residual strength and these factors were primarily expressed in the forms of empirical formulas and trend lines with domain boundaries (Lupini *et al.*, 1981; Cancelli, 1977; Kenny, 1977; Skempton, 1985; Sharma & Bore, 2003).

Field investigations suggest that slip zones of 50% of the currently identified landslides in TGP area are located at the colluvium-bedrock contact, while around 40% of the slip zones are along the bedding planes of weak rocks and the remaining 10% are characterized by the slip zones along the joints or faults across bedding planes of layered rocks (Wen *et al.*, 2007). Soils forming slip zones in the TGP area are mainly characterized by: (1) a large range of particle size (from clay to cobble); (2) abundant coarse fraction (>0.06 mm, 20–40%); (3) a wide range of Atterberg limits ($\omega_P$: 10–25%, $\omega_L$: 20–60%); (4) diversity of clay mineralogy (illite, smectite, kaolinite, chlorite); and (5) diversified properties of coarse particles (stiffness and shape). Hereinafter, we call the soil with above characteristics as composite soil. These inherent properties make such soils different from cohesive soils, which have been largely investigated for decades, and pose difficulties in investigating their residual strength, creating a gap in our understating of their mechanical behavior. This may partly be because of the difficulties in laboratory tests on such soils, which requires both a large specimen chamber and long shear distance that are not possible with common test apparatuses, and partly because of the difficulties to interpret the extremely complex and variable patterns of shearing of these soils.

For mitigating the threat of landslide failure to navigation in the TGP reservoir and to operation of the TGP dam, a better understanding of the shear behavior of these composite soils is essential and urgently needed. This can be achieved by a systematic laboratory testing and developing non-traditional analysis methods. The findings of such a study will have a much wider relevance in many applications involving composite soils.

## 1.2 SCOPE AND OBJECTIVES

This book presents a systematic and encompassing study of the shear behavior of slip zone soils with abundant coarse particles. The specific aim was to contribute to the understanding of the residual shear behavior of natural slip zone soils of landslides in the TGP area. This aim was pursued through developing a clear understanding of the influences of the following five predefined aspects/factors on the shear behavior and shear strength: (1) particle size distribution and relative percentages of each fraction; (2) Atterberg limits; (3) particle shape; (4) shearing rate; and (5) stress history and normal stress. In particular, the emphasis was put on the mechanism and process of mobilization of residual strength. In order to develop a quantitative and qualitative model for the shearing characteristics of composite soils, much effort was made to (1) interpret characteristics of residual strength, shear dilatancy and shear pattern, particularly nature of the fluctuations of measured shear stress; and (2) define the contribution of each factor and the influence of the combination of the factors, such as Atterberg limits, particle size distribution, particle shape, shearing rate, stress history and normal stress.

This book involves a set of systematic tests on artificial mixtures composed of one or more components of ideal materials, including glass sand, glass beads and clay, to establish a reference for the shear pattern of natural composite soils. A total of 138 lab tests were carried out using large ring shear apparatus and intermediate direct shear box. The normal stress ranged from 50 to 400 kPa and the shearing rate from 0.06 to 600 mm/min.

## 1.3 ORGANIZATION OF THE BOOK

The present chapter gives a brief description of the problem motivating this study and the scope and objectives of the book. An outline of the following five chapters is provided below.

Chapter 2 presents a review of relevant literatures on the mechanical behavior of soils. The review is presented in an approximately chronological order and covers all the factors that are believed to influence the mechanical behavior of soils and will be systematically investigated in this book. This chapter ends with a critical summary categorizing and briefly interpreting the main factors.

Chapter 3 is on the equipments and test techniques employed for determining the shear strength (both peak and residual) of soils. More emphasis is given to the large ring shear apparatus since it is less frequently used in the literature than the other common apparatus (e.g. direct shear box). The procedures and specifications followed for testing soil index properties and particle shape parameters are also briefly described.

Chapter 4 describes a systematical series of tests by an intermediate direct shear box (100 × 100 × 44 mm). The test materials were artificially mixed by taking one (e.g. pure glass sand or pure clay), two (e.g. glass sand + clay, glass beads + clay) or three (e.g. glass sand + glass beads + clay) components of the basic fractions (clay, glass sand and glass beads) at different weight contents. The tests produced the shear patterns of ideal materials (clay, rounded and smooth glass sand/beads) and revealed the main factors influencing the shear strength. A preliminary model for residual shear modes of composite soils was proposed based on these results. A set of artificial mixtures composed of various contents of natural materials were tested under the same conditions as the ideal samples. The results from this set of tests reinforced and expanded the model derived from the tests on ideal samples.

Chapter 5 describes the results of tests by a large ring shear apparatus (outer diameter: 310 mm, inner diameter: 210 mm and thickness: 90 mm) on samples collected from the slip zones of three landslides in TGP area. The tested samples differ from each other in terms of Atterberg limits, particle size distribution and particle shape. Each sample was tested at different shearing rates. The goal of these tests was to investigate the shear mechanisms of composite soils when they mobilize to residual condition in the large ring shear apparatus through the analysis of the influence of above factors. The model developed in Chapter 4 was improved by taking into account the ring shear testing results.

Chapter 6 presents the major conclusions of the entire investigation. Limitations and suggestions for further work on this topic are also presented on the basis of the achievements of this book.

Chapter 2

# Achievements in this field

## 2.1 INTRODUCTION

The need to understand the basic mechanisms of landslide behavior and other problems involved in the movement of soil mass in geotechnical engineering has been pushing forward the development of the knowledge of residual strength of soils. This chapter focuses on previous achievements related to the subject of the shear behavior of soils. In the first section, researches on the residual shear strength of fine-grained soils are reviewed. Following this is a detailed review of the shear behavior of soil mixtures composed of fine matrix and coarse fractions. From the point of view of the historical development of this research field, a comprehensive review of different aspects of past work is presented in an approximately chronological order within each section. In the last section of this chapter, main points emerging from previous researches are summarized by categorizing in accordance with main factors influencing shear behavior of soils.

The previous investigations on fine soils, especially clays, mainly focused on the effects caused by intrinsic properties of soils, such as Atterberg limits (liquid limit, $\omega_L$, and plasticity index, $I_P$), particle shape (platy or angular), stress history and by testing conditions, such as test method (triaxial, ring shear or direct shear), normal stress level and shearing rate. This review dates back to the first investigator, Collin (1846), who formally used the shear strength of clay specimens for the analysis of slope stability, to the first researcher, Terzaghi (1925), who laid down the foundation of soil mechanics, and to the first investigators, Gruner and Haefeli (1934), who developed the ring shear apparatus. The famous researchers worthy to mention in the early stage of this field are Collin (1846), Hvorslev (1936, 1939, 1952, 1960), Skempton (1956, 1964, 1985), De Beer (1967), Kenney (1967), Townsend (1973) and so forth.

A review of the published literature reveals that majority of research has been concentrated on pure clays or sands, however, for composite soils (gravelly soils, sandy soils or even gravel-sand-silt-clay mixtures, for instance), although the physical properties, e.g. consistency limits, compaction characteristics, dry density, and permeability have been well studied (Seed *et al.*, 1964; Day, 1989; Hsu & Saxena, 1991; Shelly & Daniel, 1993), the mechanical properties (e.g. residual strength) of such soils have received less attention. The relevant researches are reviewed and main achievements are also summarized in this chapter.

## 2.2 PREVIOUS INVESTIGATIONS

### 2.2.1 Cohesive soils

Collin's great work "Landslides in Clays" (1946) was translated into English in 1956 with an introduction "Alexandre Collin (1808–1890) and his Pioneer Work in Soil Mechanics" by Skempton (1956). In this introduction, Skempton (1956) stated that it is Collin (1846), who perhaps, is the first investigator, formally describing the characteristic rotational slip and using the shear strength of clay samples for the analysis of slope stability (Skempton, 1956). In this work, Collin (1846) in turn stated that a slip zone was, probably for the first time, reported by Vauban in 1699. From then on, the understanding of the mechanism of landslides was slowly moving forward and getting mature with the help of Coulomb's theoretical approach (Coulomb, 1773) to stability problems and the recognition of the importance of cohesion.

It is commonly known that it was Terzaghi (1925) who founded the soil mechanics as a recognized discipline, however, Hvorslev (1936), probably is the first person who put forward the concept of residual strength in the literature, following the development of shear test apparatus adopted for measuring shear strength at large displacement. It is again Hvorslev (1939) who for the first time gave a comprehensive review of early apparatus for the measurement of soil strength. In the paper "Torsion shear tests and their place in determination of the shearing resistance of soils", Hvorslev (1939) described the objectives of different shear apparatuses in detail. He illustrated that triaxial compression is the best method for determination of the maximum shearing resistance and for obtaining the stress-strain relationship and volume changes before failure, but not suitable for determination of the decrease of shearing resistance after failure. He recognized that ring shear test is the best method for determination of the decrease of shear resistance after failure for its capability allowing samples to be sheared at a large displacement and that reversal direct shear can be an alternative for this purpose.

In his Rankin Lecture, Skempton (1964) presented the results from slow, drained shear box tests indicating the post-peak decrease of shear resistance of soils. He proposed that the residual strength of a sample can be expressed by the equation $S'_r = \sigma' \times \tan\varphi'_R$, considering that the cohesion intercept of the strength envelope decreases to a value close to zero. He also introduced the residual factor $R = (S_f - \bar{S})/(S_f - S_r)$, which was used to measure the proportion of the slip zone in a clay slope along which the strength has mobilized to its residual value. Skempton suggested that the decrease of shear strength from peak to its residual value is partly because of an increase in water content within the soil matrix caused by shearing, and partly because of the orientation of clay particles which takes place upon unidirectional shearing. Another important point that emerges from the lecture is the statement that residual strength of a certain clay is unique for a particular normal effective stress level, independent of previous consolidation history. In the case where the particles involved cannot become orientated, the residual friction angle $\varphi'_R$ can be given by the equation: $\tan\varphi'_R = \frac{\pi}{2}\tan\varphi_\mu$, where $\varphi_\mu$ is the interparticle friction of clay minerals. Finally he concluded that in the general case where the soil consists of a mixture of platy and spherical particles, the residual strength would be above $\varphi_\mu$ because the presence of spherical particles would interfere in the orientation of platy particles, thus causing an increase in shear resistance.

Petley (1966) presented the results from strain-controlled reversal direct shear tests on London clay. He may be the first one to present a systematic investigation of the effects of shearing rate on residual shear. He varied the shearing rate from $4 \times 10^5$ mm/min to 1.0 mm/min, as soon as the residual conditions were reached. He concluded that the influence due to changing of shearing rate was ignorable since varying shearing rate within this range caused an increase in residual shear strength by about 4%.

Skempton and Petley (1967) described the study on the strength of overconsolidated clays along structural discontinuities. They firstly classified the discontinuities in stiff clays into three types, bedding surface, minor shear and principal displacement shear, according to the structural type and the amount of relative movement. They found that for "principal displacement shear", which had moved more than 10 cm, the residual state is easily reached at relatively small shear displacement, while in the case of "minor shear surface", which had a relative movement less than 1 cm, the strength was somewhat above the residual strength. Based on the tests on brown London clay, they indicated that the envelope of residual strength appears non-linear when under low normal stress.

De Beer (1967) presented strain controlled ring shear tests on remoulded and pre-cut specimens of Boom clay at two shearing rates of 0.0066 and 0.035 mm/min. He emphasized the influence of shearing rate on residual strength, the slower the shearing rate, the higher the residual friction angle. He explained this increase in residual strength as due to the recovery of strength in the shear zone when the soil is sheared at a significantly slow shearing rate. By comparing with the tests on specimens without pre-cut, De Beer indicated that pre-cutting specimen before consolidation leads to a reduction in residual friction angle for both of the rates used.

Kenney (1967) reported the influence of mineral composition of soils on their residual strengths, by employing reversal shear tests on natural soils and mineral mixtures. He mainly tested the influence of mineralogy, normal effective stress, pore water chemistry and shearing rate on residual friction angle. Kenney concluded that the residual friction angle of mineral mixture does not depend on plasticity, clay fraction and water content, but mainly on mineral shape. He found that rounded particles have smaller values of $\varphi'_R$ than aggregates of angular particles and that the residual coefficient of friction for natural soils is more dependent on mineral composition than on normal effective stress level and pore water chemistry. Kenney also indicated that the residual coefficient of friction for montmorillonite and hydrous mica is influenced by pore water chemistry. High concentration of cation of high valence in the pore water may cause increase in residual strength. He attributed this increase in strength to a net increase of attraction between clay particles in parallel arrangement.

Ramiah *et al.* (1970), working on a similar line of Kenney (1967), reported the results from reversal direct shear tests on a silty clay. They treated the specimens by using flocculant and dispersant, respectively before shearing. The results led them to conclude that a change from flocculative to dispersive condition of the soil causes a decrease in residual friction angle of approximately 2 to 5 degrees. The reason for this, as Ramiah *et al.* (1970) stated, is due to the higher ion concentration in the pore water and the higher interparticle contacts in flocculating condition than in dispersive state. They also mentioned that stress history and initial water content have no effect on the residual strength of tested soils. Moreover, they found that the residual stress ratio ($\tau_R/\sigma'$) for tested samples increases only slightly within a range of shearing rates between 0.025 and 10.2 mm/min.

Cullen and Donald (1971) performed reversal direct shear tests on clays to investigate influences of various parameters on residual shear strength. According to the results, they indicated that, under a given normal stress, the residual stress ratio is a function of water content at residual state. They also found that the residual strength is independent of initial placement conditions, in good agreement with what has been found by Skempton (1964), Petley (1966), La Gatta (1970), etc. Furthermore, they argued that the cohesion intercept may not always be zero.

Townsend and Gilbert (1973) measured the residual strength of various clay shales by means of reversal direct shear and ring shear. It is found that these two test methods gave the similar results in the case of hard, highly consolidated clay shales. They also concluded that stress history, initial placement conditions (intact or remoulded), initial specimen density and even normal effective stress larger than 150 kPa have no influence on residual strength of clay shales.

Morgenstern and Tchalenko (1976a) carried out microscopic observations on features at progressive stages (pre-peak, at peak and post-peak) within the shear zones of kaolin subjected to direct shear. The tests results suggested that there is no discontinuity appearing at or before peak strength; as shearing progresses, the discontinuities develop at high angles to the horizontal and extend through the middle of the sample until they eventually joint to form a wavy surface. This wavy surface would be mobilized to be planar if enough shear displacement is allowed. They also concluded that the shear-induced fabric can be explained in terms of combination of basal-plane gliding producing translation and rotation. They also mentioned the probability of a non-homogenous strength distribution on shear surface due to high stress concentrations at the loading sides of shear box. In another paper by Morgenstern and Tchalenko (1976b), they presented the results of microstructural observations on shear zones in different natural clay slips. They found that the shear zones, in general, are between 10 μ and 100 μ in thickness, but occasionally several millimeters to several centimeters.

Kenney (1977) conducted tests on mineral mixtures of massive minerals (quartz) and platy clay minerals, following his previous study in 1967. He emphasized that the residual strength of such mineral mixture depends on its composition and the chemistry of the clay mineral. The volume ratio, which is the volume of clay minerals and water divided by the total volume of sample, is found to be inversely proportional to the residual strength of whole sample.

Lemos (1986) presented a detailed investigation on the effects of shearing rate on residual shear strength of pre-established shear zones in a wide range of soils including non-cohesive sands and kinds of clays. By adopting the concept of shear modes proposed by Lupini *et al.* (1981), Lemos (1986) concluded that the residual strength of soils having low clay fraction and exhibiting turbulent shear mode is not affected by varying shearing rate; soils with high clay fraction and turbulent or transitional mode gain a decrease in residual strength with increasing shearing rate; the soils with sliding shear mode gain increase in residual strength with increasing shearing rate. In his later research, Lemos (1991) put further arguments on the behavior of soils subjected to fast shear. He noted that soils with 5% clay fraction or less are independent of rate changing; soils with 50% clay fraction or more show positive rate effect (strength increases with shearing rate); and soils with clay fraction between 5–40% exhibit negative rate effect (strength decreases with increasing shearing rate). Lemos (1991) owed the positive rate effect to the disturbance of the shear zone by massive particles and the negative rate effect to the lifted porosity and water content of shear zone.

Tika (1989) described strain-controlled ring shear tests on both natural and artificial soils with various gradations to discover the effect of shearing rate on residual strength. A shear zone was formed at slow shear and then continuously sheared at fast and slow shearing rates alternately. Tika (1989) defined three effects of fast shearing as neutral, positive and negative depending on whether the fast strength remains equal to, greater than, or less than its slow residual strength respectively. She found that all soils exhibit an increase in peak strength when resheared at fast rates. Tika (1989) argued that the positive rate effect may be attributed to viscous effects in soils exhibiting transitional or turbulent shear mode. Another possibility for this may be the disorientation of particles in soils with sliding or transitional shear mode. She took the increased porosity of shear zone and decreased viscosity of soils within shear zone as the reason causing the reduction of residual strength.

Parathiras (1994) conducted a thorough investigation of ring shear tests on both plastic and non-plastic soils to examine the influence of shearing rate on residual shear strength. Based on the comparison of fast residual strength with slow residual strength, he concluded that in the absence of an undulating shear surface, all plastic soils could gain fast residual strengths 250% higher than their slow residual strengths and all non-plastic soils could gain the fast residual strengths up to 10% lower than their slow residual strengths. When an undulating shear surface appears, all plastic soils exhibit a decrease of about 22% in fast residual strength compared to slow residual strength. The testing results indicated that the residual strength is independent of stress history and the direction of shearing, and plasticity index shows no correlation with the magnitude of the change (increase/decrease) of residual strength due to varying shearing rate.

Suzuki *et al.* (2000) suggested a linear correlation between residual strength and shearing rate in logarithm scale. Suzuki *et al.* (2001) also used Bishop-type ring shear apparatus to test clay (kaolin) and mudstone at shearing rates varying from 0.02 to 2.0 mm/min and found that the shearing rate influenced the residual strength of these materials.

Wan and Kwong (2002) conducted direct shear and ring shear tests on soils. The soils contain 55–74% amorphous materials in its clay fraction, from a slowly moving landslide in eastern Honolulu, Hawaii. The direct shear tests indicated that undisturbed soil samples have a peak cohesion of approximately 50 kPa, and internal friction angle of about 10°. They concluded that it is the amorphous clay-size material that provides strong interparticle bonds for the soils and the drop from peak strength to residual strength is mainly attributed to the breaking of these bonds and redistribution of amorphous clay-size materials. They also argued that the drained residual failure envelope is tress dependent. They owe this to the interaction of amorphous clay-size materials with slit and sand size particles: amorphous clay-size materials act as contacts between crystalline particles and therefore the contact increases with consolidation stress.

Mesri and Shahien (2003) presented an analysis of 99 cases of landslide failure in 36 types of soft clays, stiff clays and clay shales. They found that for reactivated slip zones in stiff clays and clay shales, the residual condition has been established in-situ and the mobilized shear strength is equal to the residual strength from laboratory tests (reversal direct shear and ring shear). For the first-time landslide failure in unstratified stiff clays of low plasticity, the mobilized strength is equal to the intact strength from laboratory tests. They argued that the correlation between shear strength and effective normal strength seems to be curved and there appears to be no shear strength induced at zero effective normal stress. They suggested to use the secant friction angle $[\varphi'_i]_s$, secant fully softened friction angle $[\varphi'_{fs}]_s$ and secant residual friction angle $[\varphi'_R]_s$ to fit the strength envelope since these angles were found to be function of effective normal stress.

Wesley (2003) observed that, for soils with liquid limit, $\omega_L$, below 50, there was no relationship between residual friction angle and the position on the plasticity chart. However, for soils with liquid limit values above 50, residual friction angle was highly dependent on the position in relation to the A-line on the plasticity chart. Most of the soils above the A-line have the $\varphi'_R < 10°$, while those below the A-line have much higher $\varphi'_R$. He found that for soils with $\omega_L > 50$ there was a good relationship between the drained residual friction angle, $\varphi'_R$, and the distance from the A-line, calculated from $\Delta I_P = I_P - 0.73 \times (\omega_L - 20)$.

Yoshida and Hosokawa (2004) presented staged compression-immersion-direct shear tests on crushed aggregates of mudstone. They found that immersing samples into water during compression and shear stages led to an additional compression, a reduction in mobilized shear strength and a reduction in the dilatancy during shearing at low normal stresses. Furthermore, immersion causes a significant reduction in the peak cohesion $c$ and a little change of peak friction angle $\varphi$. The fitted lines for compression at immersed condition in the plane of void ratio versus applied vertical pressure seems to parallel to those at non-immersed condition. So do the lines at critical state. They indicated that there is a gap between the parallel lines of shear stress versus void ratio, and owed this gap to the appearance of combined effect of cementation retained in a crushed mudstone aggregate itself and interlocking effect of aggregates.

Suzuki *et al.* (2005) conducted ring shear tests and reversal shear box tests on different types of cohesive soils. Together with the results reported in the literature, they produced a series of plots taking $\tan\varphi$ and the ratio of $\tan\varphi_R$ to $\tan\varphi_P$ (suffixes $R$ and $P$ denoting residual and peak state, respectively) as y-coordinates, and clay fraction, sand fraction, liquid limit, plasticity index, the ratio of plastic limit to liquid limit, activity and the molality of $Na^+$ and $Ca^{2+}$ as x-coordinates. They proposed the following relationships between residual shear strength and common soil index properties, such as clay fraction (<2 μm), $F_{clay}$, the pH value and salinity of pore water, liquid limit, $\omega_L$, and plasticity index, $I_P$.

$$\tan\varphi'_R = 3.449 I_P^{-0.737} \tag{2.1}$$

$$\tan\varphi'_R = 3.449 \omega_L^{-1.094} \tag{2.2}$$

$$\tan\varphi'_R = 1.045(\omega_P/\omega_L) - 0.077 \tag{2.3}$$

Finally, they postulated a multi-regression formula by using $F_{\text{clay}}$ and $I_P$ as the main factors:

$$\tan \varphi'_R = 10^{1.143} F_{\text{clay}}^{-0.582} I_P^{-0.531} \tag{2.4}$$

Tiwari and Marui (2004) presented results of ring shear tests on 5 specimens. They employed three test strategies, i.e. individual specimen tests under constant normal stress, increasing load multistage tests and decreasing load multistage tests. The testing results indicated that effective residual friction angle is not dependent of the test strategies used. They found that the thickness of shear zone increases with shear displacement and the residual cohesion intercept, $c'_R$, decreases with increasing shear displacement while the residual friction angle, $\varphi'_R$, remains constant.

Tiwari (2005) correlated residual friction angle with clay fraction, liquid limit, plasticity index, specific surface area and the proportion of smectite and kaolinite, and proposed that residual friction angle of soils with significant proportion of clay minerals could be estimated from a triangular correlation chart.

### 2.2.2 Composite soils

Holtz and Gibbs (1956) may be the first person to investigate the shear strength of gravelly soils, which is a mixture of gravel and sand, by means of a very large triaxial apparatus with its diameter of 230 mm. In this study, Holtz and Gibbs (1956) mainly investigated relations between shear resistance and potentially influencing parameters, such as strain rate, density, content of coarse fraction, gradation, maximum particle size, and particle shape. Based on the results of this study, they concluded: (1) there is no significant influence from the change of strain rate when it ranges from 1.8 to 0.086 percent per minute; (2) the shear strength increases with the increase of gravel content up to 50% to 60% (by weight); (3) the shear strength is significantly influenced by the shape of particles: angular quarry material causes much higher shear strength than subrounded and subangular river materials; and (4) the density is found to be a main factor to influence the shear strength.

Holtz and Willard (1956) presented an investigation into the shear strength of clayey soils with gravel contents varying from 0% to 65%. They found that increasing content of gravel would cause an increase in shear resistance and a decrease in apparent cohesion. For the mixture with more than 50% gravels, the effect of the coarse part of the mixture was readily apparent. This result was in good agreement with what Shakoor and Cook obtained in 1990 (Shakoor & Cook, 1990), when they carried out a number of unconfined compressive tests on specimens of silty clay containing varying proportions of aggregate as the coarse fraction.

Miller and Sowers (1957) studied the effects of varying proportions of sand and clay in their mixture on shear strength by means of unconsolidated undrained triaxial tests on reconstituted specimens. A series of mixtures were prepared, ranging from 100% sand to 100% clay. They found that there was no apparent change in the friction angle while the cohesion decreased gradually with increasing aggregate content. A sharp change occurred in the soil behavior for aggregate content between 67% and 74%, with the angle of shearing resistance rapidly increasing and the cohesion rapidly decreasing. This indicated that the range of gravel content (67%–74%) may be considered as a threshold, after which the granular skeleton rather than the clay matrix start to control shear behavior of the mixture. In excess of 74% gravel content, the friction angle kept increasing with the gravel content.

Smart (1970) tested the influence of clay content on the residual shear strength of soils. Based on the assumption that soil during shearing could be considered as two rigid blocks sliding one over the other, once a distinct shear surface has been established in it, Smart (1970) suggested the equation below for estimating the residual friction angle of mixtures composed of sand grains and clay particles by using the residual friction angles of the clay ($\varphi_c$) and the sand ($\varphi_s$) tested on their own:

$$\tan\varphi_R = c^2 \tan\varphi_c + 2c(1-c)\tan\varphi_{c\text{-}s} + (1-c)^2 \tan\varphi_s \tag{2.5}$$

where, $c$ = clay fraction; $\varphi_c$ = residual friction angle of clay; $\varphi_{c\text{-}s}$ = clay-sand residual friction angle; and $\varphi_s$ = residual friction angle of sand.

Patwardhan *et al.* (1970) presented the results of direct shear tests on boulders and boulder–clay mixtures. They used a large direct shear box with the diameter of 910 mm × 910 mm to meet boulders of average size of 150 mm. The boulder content varied between 0% and 100% (by weight). The results showed a gradual increase in shear strength of this type of mixture with the increase in boulder content.

Donaghe and Torrey (1979) conducted consolidated undrained triaxial compression tests on samples of gravel-sand-clay mixtures. In this study, Donaghe and Torrey (1979) used a 381 mm-diameter apparatus to test the full-scale mixtures and used a smaller 152 mm-diameter apparatus to test the specimens composed of gradations of material obtained by scalping oversized particles from the full-scale gradation and either discarding the scalped fraction or replacing it with an equal weight of finer graded material. By comparing the testing results from these two procedures, they concluded that the scalping and replacement procedure provides a satisfactory means for determining strength parameters for the mixture in terms of effective stress. The second procedure, which replaces the oversized particles with an equal weight of finer graded material, proved to produce comparable friction angle in the domain of total stress. Furthermore, the effects of specimen sizes and testing equipment proved not to influence the strength parameters significantly.

Lupini *et al.* (1981) conducted a thorough investigation on the shearing mechanism associated with residual shear strength of soils. By using ring shear apparatus, they tested the artificial mixtures of medium sand mixed with powdered mica, Happisburgh clay mixed with London clay and quart-silt mixed with sodium bentonite. They noted that for the mixture of sand and mica, the increase in normal stress may cause closer packing and interlocking of the granular particles, which in turn increase the shear strength. The result also indicated that shear strength decreases with the decrease in particle size. From the results of the observation of post-failure structures in the mixtures of Happisburgh and London clay, they postulated that for samples with a clay fraction below 30%, no distinct zone of oriented clay particles appears, and that the residual strength is close to the normally consolidated peak strength; however for the samples with a clay fraction above 35% or 40%, there is a distinct shear zones of oriented clay particles which are associated with low residual strengths. Similar results from the observation of post-failure structure were observed from the quartz-bentonite mixtures. Lupini *et al.* (1981) proposed three modes of residual shear behavior, i.e. turbulent, sliding and transitional, depending on dominant particle shape and on the coefficient of interparticle friction. The turbulent mode occurs in soils with high content of rotund particles or with platy particles with high interparticle friction. This mode produces high residual strength which depends on particle shape and packing. The sliding mode occurs in soils with a high content of platy particles of low interparticle friction. This mode is associated with low residual strengths which depend on mineralogy and pore water chemistry. The transitional mode occurs in the soils without dominant particle shape. This mode is characterized by the discontinuous appearance of sliding parts and turbulent parts along the shear zone. The residual friction angle in this mode is sensitive to the particle size distribution of the soil. Lupini *et al.* also found that the shearing mode and the residual strength have good correlation with the granular void ratio $e_g$ of the soil, which is firstly defined by Kenney (1977) as the ratio of the volume of platy particles plus voids to the volume of rotund particles:

$$e_g = \frac{\text{volume of platy particles and water}}{\text{volume of rotund particles}} \tag{2.6}$$

Vallejo (1989) put forward a model for estimating the shear strength of the mixtures composed of both fine matrix and gravel-size particles. According to his model, when the volumetric grain concentration ratio (volume of granular particles divided by volume of fine matrix) is greater than 0.8, the shear strength of the soil mixture is manly from the frictional shear resistance between large particles. When the volumetric grain concentration ratio is between 0.8 and 0.55, the shear strength of soil mixture is governed by both the frictional shear resistance among large particles and shear strength of the clayey matrix. He defined a formula for the estimation of shear strength for this type of soil mixture:

$$S = C\sigma' \tan\varphi' + (1 - C)c_\mu \tag{2.7}$$

where, $C$ = volumetric grain concentration ratio; $\varphi'$ = effective angle of internal friction between the large particles; and $c_\mu$ = undrained shear strength of the fine matrix. When the value of C is less than 0.55, the shear strength of soil mixture is only controlled by fine matrix.

Georgiannou *et al.* (1990) conducted triaxial compression tests on clayey sands (clay content <10% by weight). They recommended the granular void ratio, $GVR$ = (volume of voids + volume of clay)/(volume of granular fraction), for describing shear behavior of clayey sands. They concluded that the clayey sands become undrained brittleness in triaxial compression before reaching the phase transformation line and the value of undrained brittleness at phase transformation increases as the clay content increases from 4.6% to 10%. They also found that the undrained brittleness in triaxial compression decreases as granular void ratio decreases and the mixtures behave as non-brittleness at granular void ratio lower than 0.75.

Irfan and Tang (1992) carried out a large number of laboratory experiments on compacted specimens containing steel balls and rock fragments as the coarse fraction and fine-grained soil as the matrix to investigate the effect of coarse particles on mass shear strength of colluvial deposits and saprolites. They concluded that a low coarse fraction content of below 10% negligibly influenced the shear strength while a coarse fraction content of above 30% led to a rapid increase of shear strength.

Vallejo and Zhou (1994) reported direct shear testing results on simulated soil-rock mixtures that were developed by mixing kaolinite clay with sand. They indicated that when the sand content was less than 50%, the shear strength of whole mixture was governed by that of the clayey matrix. When the concentration of sand varied between 50% and 80%, the shear strength of whole mixture was provided in part by the shear strength of kaolin and in part by the frictional resistance between sand grains. In the case of sand concentration less than 50%, the shear strength of whole mixtures was governed entirely by that of the clay.

Vallejo and Mawby (2000) conducted a series compaction tests under three levels of static normal stresses to measure the porosity and shear strength of sand-clay mixtures. The testing results indicated that the porosity of the mixture is dependent on the fractional concentration (by weight) of sand and clay particles. And the porosity of the mixture was found to be less than that of each component, sand and clay. From the direct shear tests, it is indicated that when the sand concentration by weight was greater than 75%, the shear strength of the mixtures was governed mainly by the frictional resistance between sand grains. When the concentration of sand varied between 75% and 40%, shear strength of the mixture was given by the contribution of both sand and clay particles. When the sand concentration was less than 40%, the strength of the mixture was governed by clay only. In his following study on the mixture of gravel-size glass beads and sand-size glass beads, Vallejo (2001) got similar pattern in terms of porosity and shear strength of this type of mixture.

Ng and Chui (2003) conducted triaxial tests on loose saturated decomposed granitic soils using reconstituted samples with 21% gravels (of up to 10 mm), 53% sand, 12% silt and 14% clay by weight. He postulated that this type of composite soils behaved just like clean sands.

Wen *et al.* (2005) examined the residual strength of soils from slip zones of about 170 landslides in TGP area. They emphasized the correlation of residual strength and index properties of the slip zone soils. They found that laboratory-determined residual friction angle of gravel-free fraction of the disturbed soils from the slip zones was closely related to clay content, liquid limit and plasticity index. In contrast, in-situ residual friction angle of these soils (i.e. including coarse fraction) exhibits very weak correlations with clay content and Atterberg limits, but shows strong dependency on gravel and fines (clays and silts) contents. The ratio of gravel to fine contents is linearly related to the residual friction angle of the in-situ soils. They concluded that clay content and Atterberg limits could be used to estimate residual strength of soils finer than 2 mm, but they are not appropriate for the evaluation of residual strength of soils containing considerable amount of gravel-sized particles. For the latter, the property of gradation (particularly the ratio of gravel to fines contents) seems to be a useful indicator for this.

Cho *et al.* (2006) presented a thorough analysis on the correlation of particle shape and packing density, stiffness and strength of sands. They used sphericity, roundness and smoothness to describe the shape of sand particles. The main achievement from their study is summarized briefly as: increasing irregularity of sand particles will cause (a) an increase in both $e_{max}$ and $e_{min}$ and their difference ($e_{max} - e_{min}$); (b) a decrease in small-strain stiffness; (c) an increase in the compressibility under zero-lateral strain loading; (d) an increase in the critical state friction angle $\varphi_{cs}$; and (e) an increase in the intercept of the critical state line (CSL).

Prakasha and Chandrasekaran (2005) described static and cyclic triaxial tests on remoulded Indian marine soils having different proportions of sand and clay. They stated that inclusion of sand grains in clay leads to decrease in void ratio and increase in friction and pore pressure response therefore results in decrease in undrained shear strength. They put forward a parameter, effective void ratio ($EVR$), defined as the ratio of volume of voids to volume of effective soil fraction (sand or clay), i.e. the effective fraction is either volume of sand or volume of clay but not the sum of the two. They found that there is unique relationship between $EVR$ and $\sigma'_{vc}$ (vertical consolidation stress) in logarithm scale and therefore $EVR$ can be used to characterize the behavior of sandy-clay mixtures.

Fukuka *et al.* (2006) presented the results of ring shear tests on coarse-gained silica sands. The apparatus used in this study is DPRI ver. 7, which is equipped with a transparent shear box and a video image analysis system, for observing the formation process of shear zone in granular materials. They found that the thickness of shear zone at high speed is thinner than that at low speed. They noted that in the initial stage of shear tests, the deformation of sand layer of upper shear box appeared noticeably under drained condition and low shear speed. They defined flow-like mode and slide-like mode according to the thickness of shear zone and the difference between normalized velocity at the nearest point to the steel plate in the upper box ($V_U$) and the one at the nearest point to the rubber edge in lower box ($V_L$). Flow-like mode is characterized by thicker shear zone and the value ($V_L - V_U$) is small, in contrast, slide-like mode being with thinner shear zone and big ($V_L - V_U$) value. They concluded that high speed shear induces slide-like mode while low speed shear induces flow-like mode.

Liu *et al.* (2006) conducted ring shear, reversal direct shear and triaxial shear tests on the samples from slip zone of Xietan landslide in the TGP area. The samples had coarse particles ranging from 0% to 40%. These tests indicated that both proportion and size of gravels had an effect on the shear strength.

Simoni and Houlsby (2006) presented the results from 87 large direct shear tests on sand-gravel mixtures with various proportions of sand and gravel fractions. Subrounded and subangular gravels were added in different proportions into Leighton sand. In order to avoid the effect of confining pressure and grain breakage on the behavior of the mixtures, all tests were conducted at a low confining pressure, approximately 90 kPa, on a wide range of relative densities. Testing results indicated that even at low gravel fractions (10%–20%), when the oversize particles are considered to be floating within the sand matrix, the peak strength, constant volume strength and maximum dilatancy rate of the mixtures, are obviously higher than those for pure sand at the same density. Simoni and Houlsby (2006) noted that the relative density ($D_r$) and the reduction in minimum void ratio due to gravel addition ($e_{\mathrm{min,sand}} - e_{\mathrm{min,mixture}}$) could be used as appreciate indicators to interpret the variations in frictional and dilatant contribution to strength due to the addition of gravel particles. They suggested a simple mode for deriving $\varphi'_{\mathrm{peak}}$ and $\varphi'_{cv}$ of the mixture at low confining pressure by adopting physical and shear strength properties of sand ($e_{\min}$ and $\varphi'_{cv}$) and physical properties of the mixture ($e_{\min}$ and $D_r$). This mode is in the form:

$$\varphi'_{cv,\mathrm{mixture}} - \varphi'_{cv,\mathrm{sand}} = 18 \times (e_{\mathrm{min,sand}} - e_{\mathrm{min,mixture}}) \tag{2.8}$$

$$\varphi'_{peak} - \varphi'_{cv} = (0.8 - 1.1 \times (e_{min,sand} - e_{min,mixture})) \times \theta_{max} \quad (2.9)$$

where, the suffix peak means the peak state; *cv* means the volume constant state; $e_{min}$ = void ratio at the minimum density; and $\theta_{max}$ = is the maximum angle of dilation.

## 2.3 SUMMARY OF PREVIOUS OBSERVATIONS ON RESIDUAL STRENGTH

It is well known that the stress-deformation and stress-deformation-time behavior of soils are of importance in any problem where ground movements are of interest. By far the most widely used equation for soil strength is the Mohr-Coulomb envelope. It is stated as the basic prototype: $\tau = c + \sigma \tan \varphi$ and its altered forms corresponding to total and effective stress conditions, where $\tau$ is the shear stress on failure plane at failure; $c$ is a cohesion intercept; $\sigma$ is the normal stress on failure plane; and $\varphi$ is a friction angle. In reality, the shear resistance of a soil is a function of more than the parameters adopted in this equation. It seems to be likely dependant on the void ratio, mineral composition, stress state, stress history, temperature, strain and strain rate, and the structure of the soil (Mitchell, 1976, 1993; Lee *et al.*, 1982). Based on the limited knowledge on all parameters mentioned above, this linear relationship between normal and shear stress is taken as the approximation for describing and calculating the soil strength. For some soils, e.g. dense granular soils and heavily overconsolidated clay, the failure envelopes may be curved and typically a power law with the form $\tau = k\sigma_f^n$ may be empirically found and used in practice.

Mitchell (1976) stated that the shear resistance of a soil would be formulized as a equation in the form:

$$\tau = F(e, \varphi, C, \sigma', c', H, T, \varepsilon, \acute{\varepsilon}, S) \quad (2.10)$$

where, $e$ = void ratio; C = coefficient associated with soil composition; $H$ = coefficient associated with stress history; $T$ = temperature; $\varepsilon$ = strain; $\acute{\varepsilon}$ = rate of strain; and $S$ = coefficient associated with the soil structure. He also mentioned that "parameters in these equations may not be independent, and the functional forms of all of them are not known". As a result, a variety of types of friction angle and cohesion have been addressed to describe the behavior of soils in specified scenarios, e.g. total stress, effective stress, drained, undrained, peak strength, and residual strength etc.

Emphasis in this section is on the summary of influencing factors which have been identified in the literature including Atterberg limits, shearing rate, particle shape, particle size distribution, test method, stress history and normal stress.

### 2.3.1 Atterberg limits

The influence of index properties on the residual shear strength is the most frequent research topic in the literature. Consequently, a plenty of correlations, either quantitative or qualitative, between the residual shear strength and index properties have emerged from the literature. Haefeli (1951) postulated the brittleness parameter as the

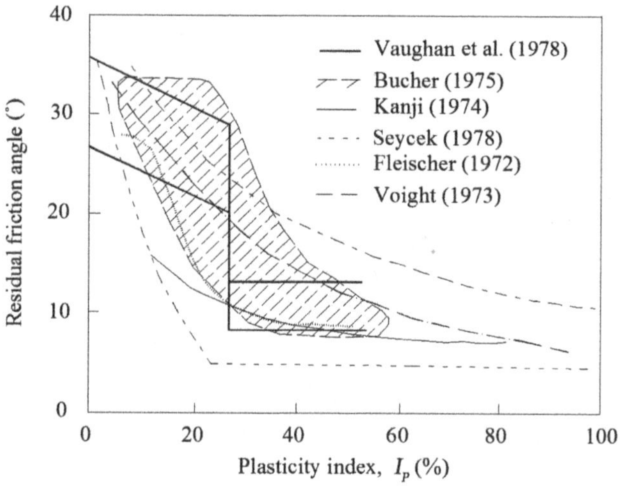


*Figure 2.1* Correlations between residual friction angle, $\varphi'_R$, and plasticity index, $I_P$ (Lupini *et al.*, 1981).

ratio of residual shear strength to peak shear strength, $\lambda_R = (\tan\varphi_R)/(\tan\varphi_P)$, where $\tan\varphi_R$ = residual shear strength; and $\tan\varphi_P$ = peak shear strength. Haefeli (1951) also realized that the magnitude of the drop in shear strength from peak to residual, $\tau_P - \tau_R$, in clays increases with liquid limit, $\omega_L$. Bishop (1967) put forward 'brittleness index', $I_B = (\tau_P - \tau_R)/\tau_P$, to quantify the difference between peak and residual strength. Correlations of the residual friction angle, $\varphi'_R$ with plasticity index, $I_P$, were also widely reported by previous researchers. Kanji (1974) postulated a relationship between $\varphi'_R$ and plasticity index by the expression: $\varphi'_R = 46.6 \times I_P^{-0.446}$ (Fig. 2.1). Fleischer and Scheffler (1979), Voight (1973), Bucher (1975), Vaughan (1975), Seycek (1978), and Vaughan *et al.* (1978) also focused on relationships between residual friction angle and plasticity index. Lupini *et al.* (1981) reviewed all the correlations obtained by these researchers and put them together in Figure 2.1. Suzuki *et al.* (2005) suggested an exponential expression for the estimation of $\varphi'_R$ using plasticity index: $\tan\varphi'_R = 3.449 \times I_P^{-0.737}$. Cancelli (1977), Mesri and Cepeda-Diaz (1986), Stark and Eid (1994) and Stark *et al.* (2005) gave the similar patterns of the relationship between $\varphi'_R$ and liquid limit (Figs. 2.2 and 2.3). Suzuki *et al.* (2005) suggested a exponential expression for the estimation of $\varphi'_R$ using liquid limit: $\tan\varphi'_R = 3.449 \times \omega_L^{-1.094}$. Townsend and Gilbert (1974) found that the residual friction angle was only dependent on clay fraction and clay mineralogy. $\varphi'_R$ and index properties were found to correlate relatively well, an expression with the general form $\tan\varphi'_R = 1 - [1/(\alpha - (\beta/c))]$ was found to describe, in the best possible way, such correlations ($\alpha$ and $\beta$ are constants and $c$ is $\omega_L$, $I_P$ or clay fraction). Collotta *et al.* (1989) introduced a correlation between the residual friction angle $\varphi'_R$ and influencing factors, such as clay fraction *CF*, liquid limit $\omega_L$ and plasticity index $I_P$, in the expression: $\varphi'_R = (CF)^2 \times \omega_L \times I_P \times 10^{-5}$.

Although the various correlations between the residual shear strength and Atterberg limits were postulated in the literature, the conflict and disagreement among

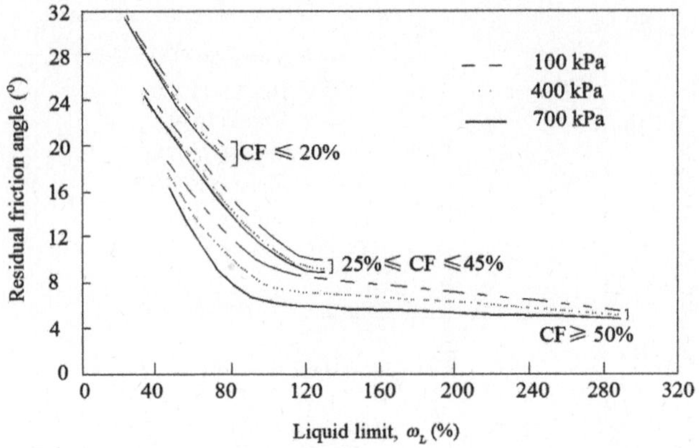


*Figure 2.2* Relationships between residual friction angle, $\varphi'_R$, and liquid limit, $\omega_L$ (Stark & Eid, 1994).

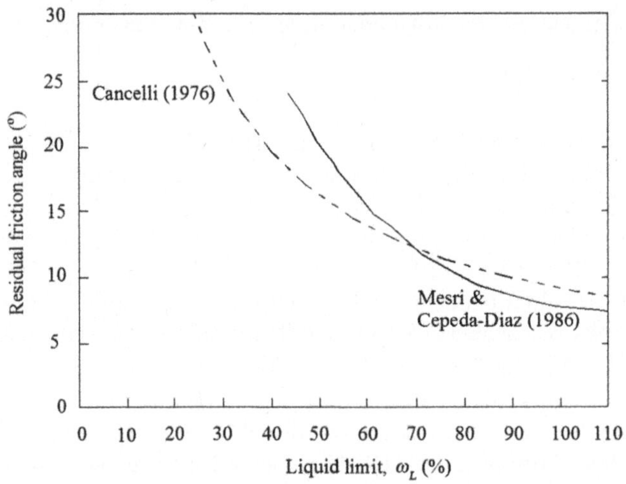


*Figure 2.3* Relationships between residual friction angle, $\varphi'_R$, and liquid limit, $\omega_L$ (Dewoolkar & Huzjak, 2005).

these estimation models are not negligible. Kenny (1967) stated that $\varphi'_R$ is not related to plasticity or grain size of the soil. Wesley (1977) stated that in the case of highly plastic residual clays, the residual shear strength did not show the same pattern as the relationship proposed by Vaughan and Walbancke (1975). Chandler (1969) also found that the index properties could not be used to relate to the residual shear strength for aggregated soils. In conclusion, although Atterberg limits have been the most popular focus when it comes to the point of residual shear strength of soils, these controversial and complicated findings from anterior researchers in the literature may guide the posterior researchers to realize that Atterberg limits must be an influencing factor to the residual shear strength but definitely not the sovereign.

### 2.3.2 Shearing rate

The reason that investigating the effect of shearing rate has been the focus of research was partly because of the need to realistically simulate the development process of slip zone and partly because of the need to accelerate commercial testing procedures and reduce the laboratory costs (Skempton & Petley, 1967; Kakou, 2001). Many investigations focused on this dependency in various soils since early 1960s (Horn & Deer, 1962; La Gatta, 1970; Idriss, 1981; Martins, 1983; Bernander *et al.*, 1985; Lemos, 1986; Tika, 1989; Parathiras, 1994). As a consequence, three standpoints at rate effects on residual strength: positive, neutral and negative effect arising from the literature.

Skempton (1985) postulated that the effect of shearing rate between 0.002 to 0.01 mm/min on residual strength was negligible in clays. Similarly, Peltey (1966) found that the varying of shearing rate from $4 \times 10^5$ mm/min to 11.0 mm/min causes an ignorable increase of the residual shear strength of brown London clay. While Yagi *et al.* (1991) indicated that residual strength of clays did not increase with shearing rate. Hungr and Morgenstern (1984) found that the shear strength of granular soils (sand and polystyrene beads) was independent of the rate of shear displacement. Sassa (1985) reported results from ring shear tests on glass beads suggesting that residual strength of granular materials were independent of the rate of displacement.

Garga (1970) found that increasing the shearing rate leads to small increase of residual strength of clay soils. Kenney (1967), La Gatta (1970) also agreed with the same point as Garga (1970) did. Herrmann and Wolfskill (1966) found the shearing rate had a small influence on residual strength, but appreciable. Cullen and Donald (1971) found that increase in shearing rate lead to an increase in residual strength of Silurian clay. Townsend and Gilbert (1974), Bucher (1975) found the shearing rate does cause increase in residual strength, especially when the rate is really high. The shearing rate of 14.6 mm/min produced a residual strength 24% higher than that of 0.0145 mm/min. Suzuki *et al.* (2000) suggested a linear correlation between residual strength and the shearing rate in logarithm scale. Suzuki *et al.* (2001) also used Bishop-type ring shear apparatus to test clay (kaolin) and mudstone at rates varying from 0.02 to 2.0 mm/min and showed that the rate of displacement significantly influenced the residual strength of these materials.

However, De Beer (1967) found that faster rate (2.1 mm/hour) causes a lower residual shear strength than slower rate (0.395 mm/hour) does for Boom clay with high liquid limit and plasticity index. Bernander *et al.* (1985) concluded that the residual strength decreases with increasing strain rate and OCR and suggested that these two factors should be taken into account when analyzing the failure mechanism of rapid landslide.

Lupini *et al.* (1981) concluded that the residual strength of soils with sliding shear mode would increase with rate of shear displacement; however the residual strength of soils with turbulent shear mode would decrease with the increase in rate of shear displacement. By means of ring shear test, Martins (1983) recognized that the fast shearing has four effects on soil strength (Fig. 2.4): (1) threshold effect associated with zero shear displacement; (2) increase in strength presumably due to restructuring of the shear zone; (3) pore pressure effect reducing the effective stresses and the shearing resistance after the slippage has occurred; and (4) non-transient increase in strength

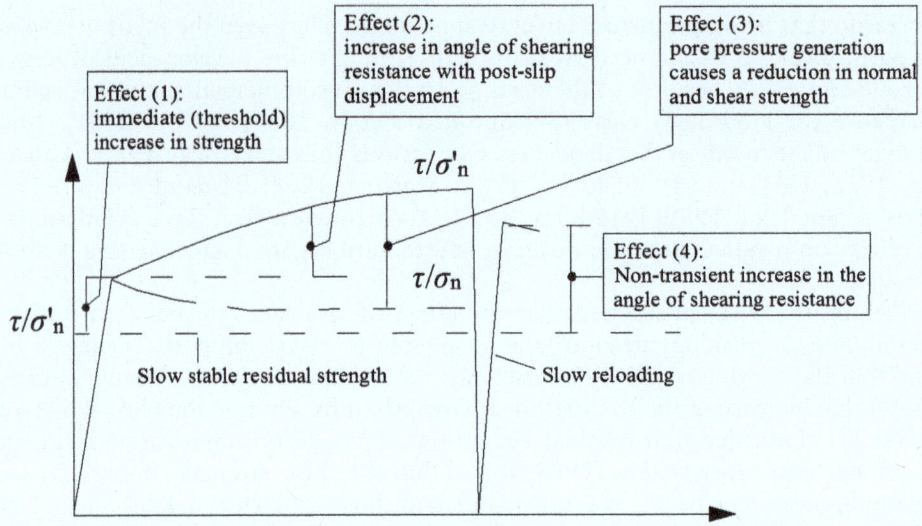


*Figure 2.4* Effects of shearing rate on shear strength (Martins, 1983).

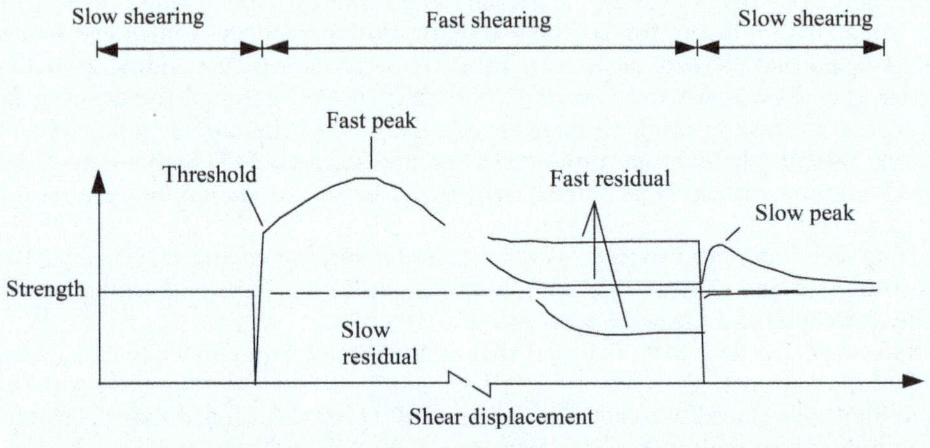


*Figure 2.5* Rate-dependent phenomena of residual strength (Tika & Hutchinson, 1999).

appearing when fast shearing is followed by slow shearing. Lemos *et al.* (1985) reported that residual strength of soils with high clay content increased and that of soils with low clay content decreased with the rate of displacement. Tika (1989) and Tika and Hutchinson (1999) concluded that there were three types of rate effects on residual strength (Fig. 2.5): positive, neutral and negative effects depending on the shear mode (sliding, transitional and turbulent), normal stress level and soil index properties (clay fraction, plasticity). Parathiras (1994) noted that the negative rate effect occurs with

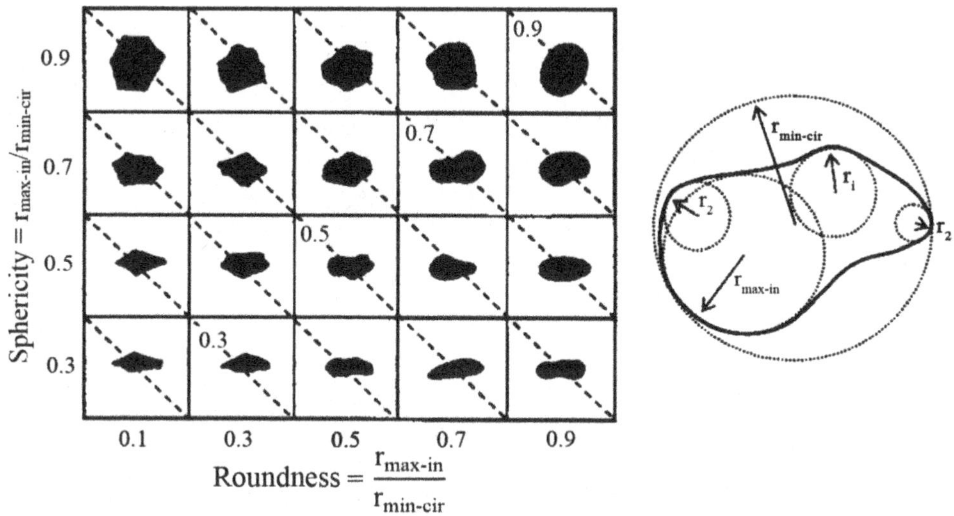


*Figure 2.6* Particle shape determination chart (Cho *et al.*, 2006).

all non-plastic soils ($I_P = 0\%$, $5\% < CF < 9\%$), while the positive effect occurs with plastic soils ($I_P > 12\%$, $\omega_L > 27\%$ and $CF > 30\%$).

In summary, the rate of shear displacement plays a role in soil shear behavior, but how and how much it contributes is mostly dependent on the soil properties (clay fraction and Atterberg limits) and stress history conditions (OCR) and may be related to the shear mode when they mobilize to residual state.

### 2.3.3 Particle shape and surface roughness

Sphericity, roundness, roughness and texture are some of the terms that were defined in many different ways and commonly used as the descriptors for the particle shape in the literature (Fig. 2.6) (Barrett, 1980; Santamarina & Cho, 2004; Cho *et al.*, 2006). The recent research revealed that increasing angularity or decreasing roundness and sphericity leads to an increase in maximum $e_{max}$ and minimum $e_{min}$ void ratios and their difference $I_e$ (Shimobe & Moroto, 1995; Miura *et al.*, 1998; Dyskin *et al.*, 2001; Jia & Williams, 2001; Nakata *et al.*, 2001; Cubrinovski & Ishihara, 2002; Santamarina & Cho, 2004; Cho *et al.*, 2006). In principal, it is recognized that the presence of platy particles in soils lead to a decrease in packing density, stiffness and residual strength, while increasing particle irregularity causes a decrease in stiffness and an increase in the internal friction angle (Fig. 2.7) (de Graff-Johnson *et al.*, 1969; Lupini *et al.*, 1981; Hight *et al.*, 1998; Guimaraes, 2002; Cho *et al.*, 2006). It is recommended by the previous researchers that particle shape emerges as a significant parameter needed to be properly characterized and documented as part of every soil characterization exercises (Alshibli & Alsaleh, 2004; Santamarina & Cho, 2004; Yan, 2005; Cho *et al.*, 2006; Li et al., 2013b; Li, 2013).

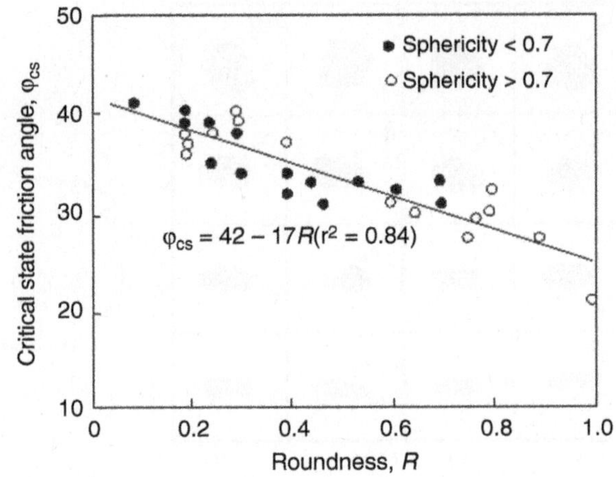


*Figure 2.7* Correlation of critical state friction angle with particle shape parameter, roundness (Cho *et al.*, 2006).

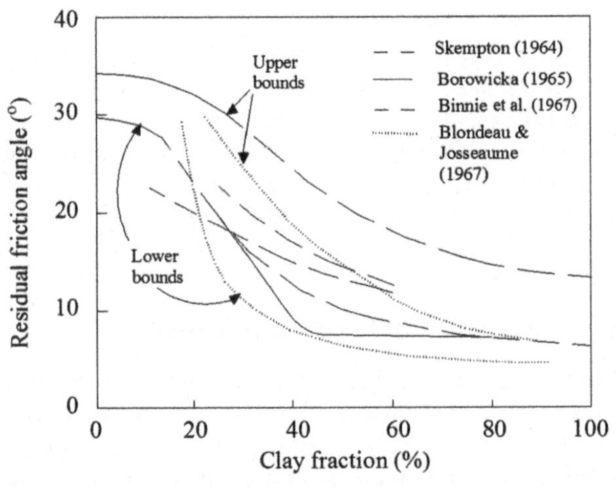


*Figure 2.8* Correlations of residual strength with content of clay fraction (Lupini *et al.*, 1981).

### 2.3.4 Particle size distribution

In the case of cohesive soils, the importance of the relative proportions of clay minerals to massive minerals (>0.002 mm in size) has been realized on several occasions (Fig. 2.8) (Skempton, 1964; Borowicka, 1965; Chandler, 1966; Vaughan & Walbancke, 1975; Lupini *et al.*, 1981). The residual shear behavior for soils containing a significant proportion of platy minerals has been associated with the re-orientation of these minerals to form a shear surface with little or no interlocking (Hvorslev, 1960; Skempton, 1964; Vaughan *et al.*, 1976, 1978; Mitchell, 1976; Lupini *et al.*, 1981). The

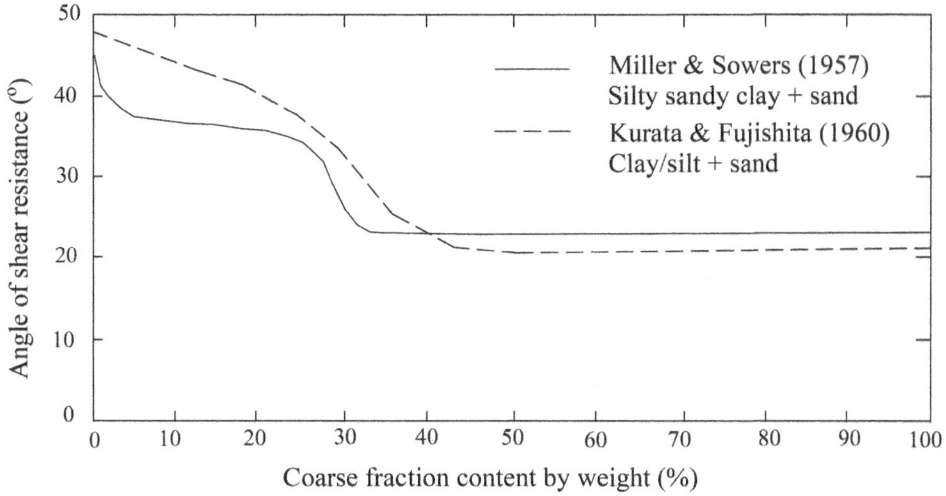


*Figure 2.9* Effect of coarse fraction on shear strength of soil-aggregate mixtures (Irfan & Tang, 1995).

residual shear behavior for soils containing a significant proportion of massive minerals have been associated with the interparticle friction and interlocking (Vaughan & Walbancke, 1975; Vaughan *et al.*, 1976, 1978; Li et al., 2012).

In the case of the soils with a wide range of particle size, often called gravelly soil, sandy soil, and even complicated as gravelly-sandy-silty-clay, or otherwise, the shear behavior is proved significantly dependent of the gradation/particle size distribution. There appears to be a nearly general agreement that a low coarse fraction content of below 10% does not influence the shear strength; a coarse fraction content of more than 30% leads to rapid increase in internal friction angle and decrease in cohesion (Figs. 2.9 and 2.10) (Miller & Sowers, 1957; Holtz, 1960; Donaghe & Torrey, 1979; Irfan & Tang, 1992, 1995; Jovicic & Coop, 1997; Ng & Chui, 2003; Liu *et al.*, 2006).

Suzuki *et al.* (2005) suggested a exponential expression using plasticity index: $\tan\varphi'_R = 10^{1.143} F_{\text{clay}}^{-0.582} I_P^{-0.531}$, for estimating the residual shear strength of soils with particle size less than 2 mm (largest size). Spear and Taylor (1972) found there was a linear correlation between residual friction angle and the logarithm of the proportion of quartz and clay minerals. Kenney (1977) proposed a function of $R_\varphi$ and the content of clay minerals and water expressed as a percentage of total volume. The relative residual strength, $R_\varphi$, was defined as:

$$R_\varphi = \frac{\tan\varphi_R - (\tan\varphi'_R)_C}{(\tan\varphi'_R)_M - (\tan\varphi'_R)_C} \tag{2.11}$$

where, $\tan\varphi'_R$ = residual friction coefficient for the mixture; $(\tan\varphi'_R)_C$ = residual friction coefficient for the clay; and $(\tan\varphi'_R)_M$ = residual friction coefficient for the massive minerals tested on their own.

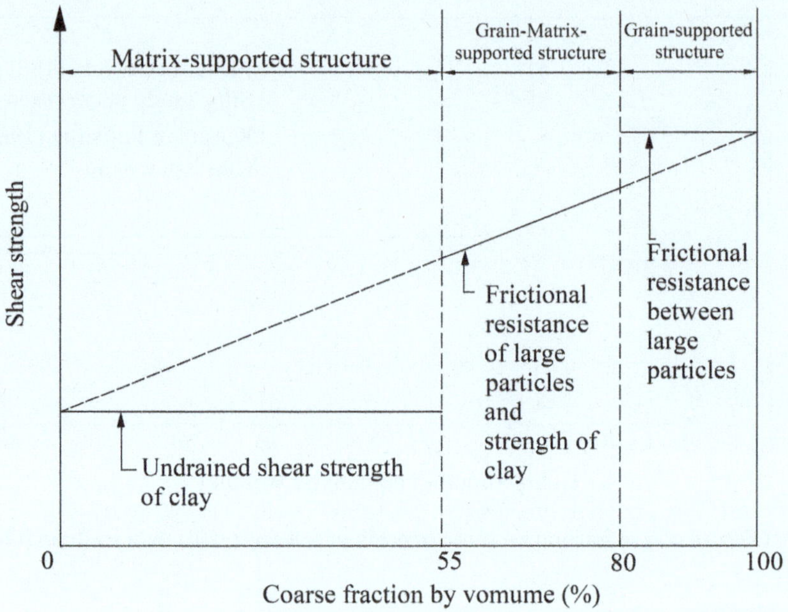


*Figure 2.10* Graphical representation of shear strength of soil with coarse fraction (Irfan & Tang, 1995).

### 2.3.5 Test method

The commonly used laboratory equipments for measuring the residual shear strength are of three basic types: reversal direct shear box, ring shear apparatus and triaxial apparatus.

In principle, reversal direct shear box is the most preferred equipment for its simplicity and general availability (Herrmann & Wolfskill, 1966; Blondeau & Josseaume, 1976; Lemos, 1986; Tika, 1989). The ring shear apparatus is proved to be the most suitable equipment to measure residual shear strength for its capability to shear the sample to unlimited displacement in one direction. The differences of the measured residual strength between these two techniques have been pointed out by some researchers (Bishop *et al.*, 1971; Hutchinson *et al.*, 1973; Bromhead, 1980). They agreed, to some extent, that the ring shear test yields lower measurements of the residual friction angle than the reversal direct shear does. Bishop *et al.* (1971) found, for London clay, the difference is about 3°. Hutchinson *et al.* (1980) pointed that for Gault clay the difference was as much as 7°. However, some investigators have found reasonable agreement between residual shear strength measured with the ring shear apparatus and the reversal direct shear box (Townsend & Gilbert, 1976; Bucher, 1975). There seems to be a common agreement that interpreting the curve of stress against shear displacement from the reversal direct shear test is more difficult than that from the ring shear test (Kenney, 1967; Bishop *et al.*, 1971).

Triaxial apparatus was generally considered not suitable for measuring the residual strength of soils for its strain limitation (Bishop *et al.*, 1965, 1971; Bucher 1975). It is found, however, triaxial could be used to test the discontinuities, such as fissured clay, pre-cut specimens and actual shear surface (Petley, 1966; Skempton & Petley, 1967). Bishop *et al.* (1965) measured the residual strength of London clay using triaxial apparatus and concluded that measurements of residual strength using triaxial compression with limited displacement led to an overestimation compared with Skempton's measurements by using direct shear box (Skempton, 1964). Hutchinson, Somerville and Petley (1973) presented results of an investigation of a pre-existing landslide. Residual strength parameters were obtained by different means. They carried out shear box tests, triaxial tests and ring shear tests on the natural shear surface. It was found that the strength from triaxial and direct shear box tended to overestimate the strength from back analysis, while the ring shear test tended to underestimate it.

### 2.3.6 Placement conditions, stress history and normal stress

Many investigators agreed that placement conditions (undisturbed, remoulded, slurried and pre-cut) do not affect the measured residual strength (Skempton, 1964; Petley, 1966; Herrmann & Wolfskill, 1966; Kenney, 1967; Garga, 1970; La Gatta 1970; Bishop *et al.*, 1971; Early & Skempton, 1972; Bucher, 1975; Townsend & Gilbert, 1976; Lupini *et al.*, 1981). There is an exception that De Beer (1967) found that pre-cutting sample before shearing gains lower residual friction angle, however many researchers argued that this conclusion is somehow weak since the residual conditions were considered not fully established. It is also a general consensus that residual strength is basically independent of loading sequence and previous stress history (Skempton, 1964; Petley, 1966; Kenney, 1967; Garga, 1970; La Gatta 1970; Bishop *et al.*, 1971; Townsend & Gilbert, 1976; Lupini *et al.*, 1981).

The normal stress, when less than 200kPa, does not seem to affect the residual friction angle, $\varphi'_R$, of most cohesive soils (Herrmann & Wolfskill, 1966; Garga, 1970; Bishop *et al.*, 1971; Chattopadhyay, 1972; Bucher, 1975; Kenney, 1976; Lupini *et al.*, 1981; Li et al., 2013c). However, for brown London clay, Weald clay, micaceous soils, the residual strength were found to show dependency on normal effective stress levels (Townsend & Gilbert, 1976). Chattopadhyay (1972) put forward a relationship between normal stress and the residual friction angle by employing elastic junction theory, where the interparticle friction angle is related to the effective normal stress $\sigma'_n$ by the expression: $\tan\varphi_\mu = \tau_i K(\sigma'_n)^{-1/3}$. Thus, in the case of platy particles, for which the residual friction angle, $\varphi'_R$, can be substituted by interparticle friction angle, $\varphi'_\mu$, this correlation can be used for the estimation of residual strength. Mesri and Abdel-Ghaffar (1993) proposed the empirical equation for the intact strength envelope by an complicated formula: $s(r) = \sigma'_n \tan[\varphi'_R]_S^P \left[\frac{\sigma'_p}{\sigma'_n}\right]^{1-m_r}$, where $[\varphi'_R]_s^P$ = secant fully softened friction angle at $\sigma'_n = \sigma'_p$; $\sigma'_n$ = effective normal stress; $\sigma'_p$ = preconsolidated pressure; and $1 - m_r$ = slope of $\log(\tan[\varphi'_R]_s/\tan[\varphi'_R]_s^P)$ versus $\log(\sigma'_p/\sigma'_n)$.

Several investigators presented that the necessary displacement to reach the residual state of soils would be related to the level of normal stress, though no general conclusion has been made. When testing Pierre clay shales by using reversal direct shear box, Herrmann and Wolfskill (1966) found that the displacement to residual state seems to be shortened with increasing normal strength. La Gatta (1970) obtained the similar conclusion for tests on remoulded Pepper shale. However, Peck (1967) conducted direct shear tests on undisturbed Lake Agassiz clay and concluded the opposite results to Herrmann and Wolfskill (1966).

Chapter 3

# Testing method

## 3.1 INTRODUCTION

The most commonly used apparatuses for evaluating the shear behavior of soil are small direct shear box and triaxial apparatus because of the simplicity of their construction and operation (Skempton, 1985). However, these apparatuses are not suitable to determine residual shear strength of slip zone soils containing abundant coarse particles and to simulate the actual landslide conditions due to their limited capacities in respect of shear displacement and large particle size of the specimen. Although the ring shear apparatus is not easy to construct and operate (Bishop *et al.*, 1971), it does have the obvious advantages that the shear displacement can be unlimited without having to stop (Bishop, 1971; Bromhead, 1979; Sassa, 1996). Alternatively, reversal direct shear with intermediate and large shear boxes can also be employed to measure the residual strength of the soils.

Three basic types of ring shear apparatuses and their modifications have been widely used: (1) IC/NGI type, developed jointly by Imperial College of Science and Technology, University of London, and Norwegian Geotechnical Institute in 1971, with the specimen chamber dimensions of OD: 152.4 mm, ID: 101.6 mm and thickness: 19 mm; used to perform drained tests under maximum normal stress of 980 kPa and a maximum nominal shear stress of 490 kPa; (2) Bromhead type, developed by Bromhead in 1979 with the specimen chamber dimensions of OD: 100 mm, ID: 70 mm and thickness: 5 mm; and (3) DPRI type, developed by the Disaster Prevention and Research Institute (DPRI), Kyoto University of Japan. The first two ring shear apparatuses are designed to test soils with fine particles (due to the limited dimensions of their specimen chambers), while the DPRI type is adequate for soils with a large range of particle size distribution.

There are three types of shear boxes commonly used to test shear strength of soils. A small direct shear box, with the dimension of 60 × 60 × 20 mm, is suitable for soils containing particles up to 4 mm in size; an intermediate direct shear box, with the dimension of 100 × 100 × 44 mm, is suitable for soils containing particles up to 10 mm; a large shear box, with the dimension of 300 × 300 × 150 mm, is suitable for soils containing particles up to 20 mm in size.

In the present study, the shear behavior of slip zone soils has been investigated by means of a DPRI ver.3 large ring shear apparatus and an intermediate direct and residual shear apparatus of ELE™. In this chapter, a brief description of the apparatuses and their peripheral systems are presented. The sample preparation techniques and the testing procedures are also described, followed by the procedures/standards used for soil classification tests (Atterberg limits, particle size distribution and specific gravity) and the measurement of particle shape.

## 3.2 DPRI TYPE RING SHEAR APPARATUS

### 3.2.1 Brief description of the apparatus

The design and procedures of operation of the DPRI ver.3 ring shear apparatus have been described in detail by Sassa (1992, 1996) and Sassa *et al.* (2004). The mechanism and concept of this ring shear apparatus have the same principle as the ring shear apparatus of IC/NGI type developed by Bishop *et al.* (1971). While the former is aimed to simulate the formation of the shear zone and the post-failure mobility of high speed landslides and to observe the mobilized shear resistance, as well as the post-failure shear displacement and generated pore water pressure; the latter is to study the post-peak interval of the shear resistance-displacement curve with emphasis on residual strength developed in slow moving clayey landslides.

The key features of DPRI-ver.3 ring shear apparatus (Fig. 3.1) developed in 1992, which was employed in this study, are listed in Table 3.1.

*Figure 3.1* The assembly of DPRI ver.3 ring shear apparatus.

*Table 3.1* Key features of DPRI-ver.3 ring shear apparatus used in this study.

| *Parameter* | *Value* |
|---|---|
| Inner-diameter of specimen chamber (mm) | 210 |
| Outer-diameter of specimen chamber (mm) | 310 |
| Maximum normal stress (kPa) | 500 |
| Area of specimen (mm$^2$) | 408400 |
| Maximum height of specimen (mm) | 90 |
| Maximum shearing speed (mm/sec) | 300 |
| Resolution of gap control (mm) | 0.001 |
| Maximum data acquisition rate (readings/sec) | 12 |
| Suitability for undrained test? | Yes |

The shear box is assembled with two pairs of confining (inner and outer) rings forming upper and lower halves. The lower half of the shear box is rotatable and driven by a servomotor, while the upper confining rings are fixed with the cap plate which in turn is fixed by a rigid arm. The normal stress is applied on the donut-shaped specimen through an annular loading platen by a pneumatic piston. Several metal teeth on the annular loading platen and on the base of the shear box prevent the relative slip along the ends of the sample.

In order to prevent leakage of water and extrusion of specimen during shear through the tight interface between the upper and lower rings, a pair of stair-shaped rubber rings with high surface smoothness is placed in the specially made grooves along the perimeters of the lower inner and outer confining rings. The contact pressure between the rubber rings and the upper pair of the confining rings is adjusted at the start and maintained throughout the test at a level greater than the pore water pressure that can possibly be generated in the specimen and as constant as possible by means of a hydraulic piston controlled by feed-back signals from a gap sensor accurate to 0.001 mm. Similarly, O-rings (inserted into special slots along the inner and outer perimeters of the annular loading platen) are utilized to prevent upward leakage. High quality vacuum silicon grease is applied on to the rubber edges to lubricate and seal them.

### 3.2.2 Measurement devices and data logging system

Figure 3.1 presents an overall view of the assembly of DPRI ver.3 ring shear apparatus. It consists of a main shear device, a set of transducers, a data logging unit, two computers, a multi-pen plotter and a printer.

The shear loads, applied through the loading yoke, are measured by a load cell mounted on a rigid stand of the stationary frame. With the initiation of the shearing, the loading yoke becomes tangential to the load cell, compresses it and causes its horizontal deflection. This deflection is then measured by means of inductive displacement transducers connected to modulator-demodulator cards with a D/C output.

DPRI-ver.3 ring shear system has two load transducers to monitor normal stresses, one in between the upper loading plate and pneumatic piston to measure the normal load applied onto the specimen, and the other to measure the sum of: (1) the vertical friction between the sample and the sidewalls of the upper pair of rings; and (2) the self-weight of the upper pair of rings. The difference of the measurements by these two transducers gives the actual normal stress acting on the shear surface. The value of this difference is sent to a servo-amplifier as a feed-back signal to keep the normal stress on the shear surface constant through automatic adjustment of the load from the pneumatic piston.

A gutter, located 2 mm above the shear surface and is covered by two metal filters, extends along the entire circumference of the inner wall of the outer ring in the upper box. Two pore water pressure transducers are connected to this gutter to measure the pore water pressure near the shear plane in the specimen.

The vertical deformation of the specimen during shearing is measured by means of a linear strain conversion displacement transducer with a maximum travel of 25 mm, which produces a proportional voltage output in relation to its spindle displacement. The relative rotational displacement (shear displacement) is measured by means of a rotational potentiometer which also produces a proportional voltage output.

The DPRI ver.3 ring shear apparatus is equipped with an electronic measuring device. All the signals from these transducers are monitored by a data logging unit, which has a measurement frequency of 12 recordings per second. A PC takes this data logging unit as an interface to connect to the transducers to collect the signals, transfer them to measurements and store into the memory.

### 3.2.3 Testing procedure

The sample was formed by means of dry deposition. Each of the air-dried and well mixed samples was poured into the shear box in layers. Each layer was tamped slightly so as to achieve the field density of the specimen. After the dry specimen was prepared, carbon dioxide ($CO_2$) was injected into the chamber very slowly through the lower drainage line to drive out the air in the voids of the specimen through the upper drainage line. This replacement of air with $CO_2$ usually took approximately 6 hours until concentration of discharged $CO_2$ passing through the specimen is adequately high to make sure that all the voids are filled with $CO_2$.

Next, through the lower drainage line, de-aired and de-ionized water was infiltrated into the specimen at a low flow rate to avoid the possible influence on specimen homogeneity by the water flow. For checking the degree of saturation, the parameter $B_D$ (parameter of saturation in the direct shear state) was employed, as proposed by Sassa (1988). This parameter is given by $B_D = \Delta\mu/\Delta\sigma$, where $\Delta\mu$ and $\Delta\sigma$ are the increments of pore water pressure and total normal stress, respectively. For the specimens tested in this study, it usually took approximately 24 hours to reach $B_D \geq 95\%$.

All the slip zone soils sampled for the purposes of this study were normally consolidated in situ. Therefore, after the saturation stage, it generally took approximately 18 hours to restore this level of in situ state of consolidation. After completion of the consolidation stage, the physical properties, such as void ratio and water content, of each consolidated non-sheared specimen were determined, and the specimen was then subjected to shearing at a prescribed shearing rate.

### 3.2.4 Calculations

Bishop (1971) suggested that the oversimplification that the normal and shear stress are uniformly distributed across the shear plane would introduce an error which is considered to be insignificant. Based on this assumption, the average shear and normal stresses are given by the equations:

$$\sigma_n = \frac{W_n}{\pi \cdot (r_2^2 - r_1^2)} \tag{3.1}$$

$$\tau = \frac{3M}{2\pi \cdot (r_2^3 - r_1^3)} \tag{3.2}$$

$$\tan\varphi = \frac{\tau}{\sigma_n} = \frac{3M \cdot (r_1 + r_2)}{2W_n \cdot (r_1^2 + r_1 r_2 + r_2^2)} \tag{3.3}$$

$$W_n = N_1 - N_2 \tag{3.4}$$

$$M = F \times l \tag{3.5}$$

where,
$W_n$ = net normal load on the sample
$M$ = torsional moment
$r_1, r_2$ = internal and external sample radii (155 mm, 105 mm)
$N_1$ = load by the pneumatic pump on the loading platen
$N_2$ = sum of side friction and self-weight of upper ring
$F$ = measurements of the load cells for shear force
$l$ = arm of shear force (390 mm)

Thus, by substituting the corresponding values of $r_1$, $r_2$ and $l$ for this test system, the stress ratio can be expressed by:

$$\tan\varphi = \frac{\tau}{\sigma_n} = 2.9634 \times \frac{F}{N_1 - N_2} \tag{3.6}$$

## 3.3 DIRECT SHEAR APPARATUS

### 3.3.1 Large direct shear machine

The S2450 large direct shear machine (Fig. 3.2) has a pair of shear boxes of rounded corners of 300 × 300 mm to eliminate crowding and stress concentration in sharp corners. The moving half of the shear box is seated on a rigid sled with both contact surfaces covered with low-friction teflon. Shear force is applied by hydraulic cylinders. The oil pressure is set and maintained by applying the output of an electro-pneumatic pressure regulator to the pressure tank. Strain rate, which has the range from 0.001 to 7.62 mm/min, is controlled by a constant pressure difference across the micrometer valves.

Vertical and horizontal displacements are measured by digital indicators with outputs for data acquisition. Four load cells measure normal load to an accuracy

*Figure 3.2* Geotest large direct shear box.

of 1 lbf. Other measured values are displayed on panel meters, also with outputs for data acquisition. Data acquisition is in real time and features both graphical and digital display of load/deformation and consolidation curves with data storage in ASCII format.

Side friction is measured by suspending (floating) the upper (stationary) half of the shear box on load cells using threaded rods. By deducting side friction from applied normal load, the stress is accurately computed at the shear-friction zone; only the true force is measured as all friction associated with the moving half is excluded.

### 3.3.2 Intermediate direct shear box

The ELE type direct and residual shear box (Fig. 3.3) consists of a micro processor control unit, a large on-board LCD screen display and measuring devices. It is equipped with two digital dial gauges (range of 0 to 15 mm, resolution of 0.002 mm), one is measuring vertical displacement and the other measuring the horizontal shear displacement. Shear force was measured by means of a proving ring with a digital dial gauge inside. All the data were recorded by means of a data transfer unit. Normal stress is applied to the specimen through the lever loading device. This apparatus accepts specimens up to 100 mm × 100 mm on plan. The rate of shear displacement varies from $10^{-5}$ to 10 mm/min.

### 3.3.3 Testing procedures

Reversal direct shear tests by means of large direct shear machine and intermediate direct shear box were conducted according to Part 7 of the British standard 1377 (BSI, 1990).

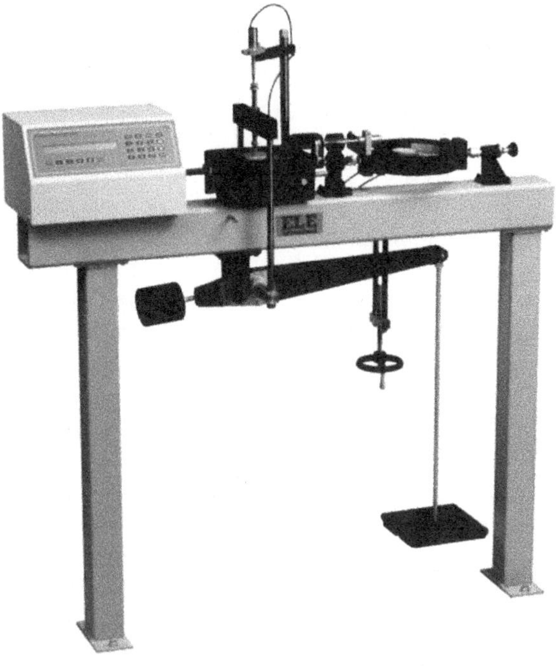

*Figure 3.3* ELE direct and residual shear apparatus.

## 3.4 SOIL CLASSIFICATION TEST

### 3.4.1 Particle size distribution

The particle size distribution was determined by wet sieving and the hydrometer method according to Part 2(9) of the British standard 1377 (BSI, 1990). Where necessary, the laser particle size analyzer was used for testing the fraction of the soils with <1 mm particles.

### 3.4.2 Specific gravity

The specific gravity of fine grained soils was determined by small pycnometers and that of coarse-grained soils was determined by large pycnometers according to Part 2(8.3) and Part 2(8.4) of the British standard 1377, respectively (BSI, 1990).

### 3.4.3 Index properties

The index properties were determined according to BS 1377:1990. The plastic limit was determined according to Part 2(5) of the British standard 1377 (BSI, 1990) and the liquid limit was determined with a cone penetrometer according to Part 2(4) of the British standard 1377 (BSI, 1990).

## 3.5 MEASUREMENT OF PARTICLE SHAPE

A stereomicroscope (Olympus SZH10) and an image capture and analysis system, analySIS$^{TM}$, were employed to quantify the shape parameters of soil particles. The stereomicroscope with its magnification capability of up to 140X enables capturing of the shape features of even fine silt fraction down to 0.002 mm. Referring to the 2D nature of the images (Fig. 2.6), the analysis software uses the term "circularity" to quantify the "closeness" of a particle to a circle. Circularity is given by $4\pi A/P^2$, where $A$ = particle area; and $P$ = perimeter. Thus a perfect circular/spherical particle has a circularity of 1 while a very "spiky" or irregular particle has a circularity value closer to 0. The term "convexity" is used as a measure of surface roughness of a particle and is given by the ratio of the "convex hull perimeter" by the actual particle perimeter. Convexity also ranges from 0 for a smooth shape to 1 for a spiky shape. The roundness of particles is expressed by elongation quantified by "1 – *width*/*length*". Particles with equal lengths along all axes have an elongation value of 0 while those with one dimension significantly larger than the other have elongation values closer to 1.

The samples were sieved by following the same procedure as for particle size analysis to separate them into several fractions. Shape parameters of each fraction were then measured as described above. The number of randomly selected particles was large enough such that a statistical average value for each parameter was achieved with variance of less than 0.01. Shape parameters for each sample were obtained by the weighted sum of the mean parameter values for each fraction (i.e. scaled by their corresponding weight percentages).

## 3.6 SUMMARY

The technical specifications for the ring shear apparatuses of DPRI type, and the direct shear boxes of ELE and Geotest types, together with their instrumentations and data recording systems have been described in this chapter as the laboratory systems which have been used for the investigations conducted in this study. The specimen preparation, testing procedures and the calculations to be carried out before the analysis of the testing results were also presented for each of these shear test apparatuses. Moreover, the soil classification tests and the measurement methods of the particle shape parameters adopted in the preparation and characterization of the sample mixtures have been described.

Chapter 4

# Artificial soil samples

## 4.1 INTRODUCTION

The work reported in the literature (see Chapter 2) reveals that the main factors influencing soil residual shear behavior are Atterberg limits, particle size distribution, normal stress, particle shape and shearing rate. The purpose of this chapter is to investigate the extent to which these factors influence the residual shear behavior of soils. This investigation was conducted by means of an intermediate direct shear box. The samples are divided into three groups of mixtures in terms of their composition: (1) a mixture of kaolin, glass sand and glass beads; (2) kaolin, silt, glass sand and glass beads; and (3) kaolin, silt, river sand and crushed granite gravels. These mixtures represent a wide range of natural soils, particularly in terms of particle size and particle shape.

A description of the samples and the testing procedures is given in Section 4.2. Testing results are presented and discussed in a broader context (including soil compressibility, the structure and water content of the shear zone) in Section 4.3. With the aid of the Fast Fourier Transform (FFT), the shear mode of granular materials is examined in detail based on the tests with rounded glass materials. The shear behaviors of composite soils are also discussed in detail based on the observation of shear patterns (development of shear stress and vertical displacement with shear displacement) and the structures of shear zones. The residual shear modes, which were put forward by Lupini *et al.* (1981) for fine-grained soils, are modified and extended to be applicable to soils with a wide range of characteristics in terms of particle size and particle shape. A quantitative analysis of the influence of the controlling factors is presented in Section 4.4. In the last section, the main conclusions derived from these tests are presented.

## 4.2 TESTING PROGRAM

### 4.2.1 Specimen characteristics

Six types of materials were chosen as constituent materials to prepare the samples. The basic properties of these materials are listed in Table 4.1. The kaolin, nonporous soda lime glass sand and glass beads are commercially available while the silt is from a natural source, the natural sand is from a river deposit and gravels (with particle

*Table 4.1* Properties of the constituent materials.

| *Soil type* | *$\omega_L$ (%)* | *$\omega_P$ (%)* | *$I_P$ (%)* | *PS (mm)* | *$G_S$ (g/cm$^3$)* | *Particle shape* |
|---|---|---|---|---|---|---|
| Kaolin | 50.3 | 22.0 | 28.3 | ≤0.002 | 2.68 | – |
| Silt | 41.9 | 18.7 | 23.2 | 0.002–0.063 | 2.65 | sub-rounded |
| Glass sand | – | – | – | 0.063–2.0 | 2.48 | rounded |
| Glass beads | – | – | – | 2.0–8.0 | 2.48 | rounded |
| Natural sand | – | – | – | 0.063–2.0 | 2.63 | sub-angular |
| Gravels | – | – | – | 2.0–8.0 | 2.59 | angular |

Note: PS: Particle size range, Gs: Specific gravity of soil particle.

size <10 mm) are crushed granite fragments. Particle size of kaolin mainly is ≤0.002 mm and that of silt ranges from 0.002 to 0.063 mm. The particle sizes of glass sand and glass beads show even distributions between 0.063 and 2.0 mm and between 2.0 and 8.0 mm, respectively. Specific gravities of these constituent materials were determined by a small pycnometer and a gas jar for applicable size of materials in accordance with BS 1377 (BSI, 1990).

Artificially mixed samples fall into three categories with respect to their constituents, the way they were mixed and the testing conditions they were subjected to. The index properties, the particle size distribution parameters and the sequence of shear tests performed are summarized in Table 4.2. Particle size distribution has been determined by means of wet sieve for the particles larger than 0.063 mm and by sedimentation for the finer particles; the particle size determined from Stokes's Law has been corrected by the appropriate factor to give the equivalent sieve grading (Allen, 1990).

The first category includes Samples 1 to 11, among which the first three had kaolin, glass sand and glass beads as constituent materials, respectively. Sample 4 was a binary granular mixture of glass sand and glass beads each weighting 50% of the total mass of the sample. Samples 5 and 6 were binary mixtures of 50% kaolin and 50% glass beads and 50% kaolin and 50% glass sand respectively. Samples 7 to 11 were ternary mixtures with varying weight proportions (80% to 20%) of kaolin and coarse fraction (20% to 80%), which was composed of 50% glass sand and 50% glass beads. Figure 4.1 presents the particle size distributions of the samples within this category.

The second category includes Samples 12 to 18. These were mixtures with decreasing weight proportions of fine fraction at intervals of 10% and correspondingly increasing coarse fraction. The fine fraction was composed of 25% kaolin and 75% silt while the coarse fraction was composed of 50% glass sand and 50% glass beads as that for the first category. The particle size distributions are shown in Figure 4.2.

The third category refers to Samples 19 to 27. These mixtures were prepared by decreasing fine fraction at intervals of 10% and correspondingly increasing coarse fraction. The fine fraction was composed of 25% kaolin and 75% silt, while the coarse fraction was composed of 50% natural sand and 50% natural gravel. The gradation of particle size of the samples in this category refers to Figure 4.3.

*Table 4.2* Physical characteristics of tested samples and testing conditions.

| *Sample no.* | | | *Sample reference** | *Clay (%)* | *Coarse (%)* | $\omega_L$ *(%)* | $\omega_P$ *(%)* | $I_P$ *(%)* | $D_{50}$ *(mm)* | $C_c$ | $C_u$ | *BSCS symbol* | *Testing conditions#* |
|---|---|---|---|---|---|---|---|---|---|---|---|---|---|
| The first category (KG mixture) | Single fraction | 1 | 100K | 100.0 | 0.0 | 50.3 | 22.0 | 28.3 | 0.001 | 1.0 | 2 | CH | I, II, III, IV, V |
| | | 2 | 100GS | 0.0 | 100.0 | – | – | – | 0.320 | 0.8 | 6 | SP | I, II, III, IV, V, VI |
| | | 3 | 100GB | 0.0 | 100.0 | – | – | – | 4.200 | 0.9 | 2 | GP | I, II, III, IV, V, VI |
| | Binary fractions | 4 | 50GS50GB | 0.0 | 100.0 | – | – | – | 2.000 | 0.8 | 26 | SPG-GPS | I, II, III, IV, V |
| | | 5 | 50K50GB | 50.0 | 50.0 | 49.9 | 21.7 | 28.2 | 0.002 | 0.0 | 157 | CIG | I, II, III, IV, V |
| | | 6 | 50K50GS | 50.0 | 50.0 | 27.4 | 12.8 | 14.6 | 0.800 | 0.0 | 2857 | MLS | I, II, III, IV, V |
| | Ternary fractions | 7 | 80K20GC | 80.0 | 20.0 | 43.8 | 19.8 | 24.0 | 0.001 | 1.0 | 3 | CI | I, II, III, IV, V |
| | | 8 | 60K40GC | 60.0 | 40.0 | 33.8 | 16.3 | 17.5 | 0.002 | 1.0 | 3 | CLS-CLG | I, II, III, IV, V |
| | | 9 | 50K50GC | 50.0 | 50.0 | 27.2 | 14.9 | 12.3 | 0.002 | 0.0 | 306 | MLS-MLG | I, II, III, IV, V |
| | | 10 | 40K60GC | 40.0 | 50.0 | 22.8 | 13.5 | 9.3 | 0.200 | 0.0 | 789 | MLS-MLG | I, II, III, IV, V |
| | | 11 | 20K80GC | 20.0 | 80.0 | 13.4 | 11.1 | 2.3 | 0.800 | 7.5 | 2083 | SPG-GPS | I, II, III, IV, V |
| The second category (FG mixture) | | 12 | 80F20GC | 20.0 | 20.0 | 39.2 | 17.9 | 21.3 | 0.009 | 0.8 | 11 | CI | I, II, III, IV, V, VI |
| | | 13 | 60F40GC | 15.0 | 40.0 | 31.4 | 13.2 | 18.2 | 0.012 | 0.4 | 43 | CLS-CLG | I, II, III, IV, VI |
| | | 14 | 50F50GC | 12.5 | 50.0 | 24.9 | 10.6 | 14.3 | 0.060 | 0.2 | 124 | MLS-MLG | I, II, III, IV, VI |
| | | 15 | 40F60GC | 10.0 | 60.0 | 20.8 | 9.1 | 11.7 | 0.205 | 0.2 | 350 | MLS-MLG | I, II, III, IV, VI |
| | | 16 | 20F80GC | 5.0 | 80.0 | 10.1 | 5.9 | 4.2 | 1.000 | 1.6 | 227 | SWG | I, II, III, IV, VI |
| | | 17 | 70F30GC | 17.5 | 30.0 | 35.7 | 15.5 | 20.2 | 0.013 | 0.5 | 22 | CI | III, VI |
| | | 18 | 30F70GC | 7.5 | 70.0 | 16.3 | 7.4 | 8.9 | 0.510 | 1.0 | 400 | SWG | III, VI |
| The third category (FN mixture) | | 19 | 80F20NC | 20.0 | 20.0 | 41.6 | 19.5 | 22.1 | 0.009 | 0.8 | 11 | CI | III, VI |
| | | 20 | 70F30NC | 17.5 | 30.0 | 37.8 | 17.7 | 20.1 | 0.013 | 0.5 | 22 | CI | III, VI |
| | | 21 | 60F40NC | 15.0 | 40.0 | 32.1 | 15.4 | 16.7 | 0.012 | 0.4 | 43 | CLS-CLG | III, VI |
| | | 22 | 50F50NC | 12.5 | 50.0 | 26.3 | 13.6 | 12.7 | 0.060 | 0.2 | 124 | MLS-MLG | III, VI |
| | | 23 | 40F60NC | 10.0 | 60.0 | 22.9 | 11.0 | 11.9 | 0.205 | 0.2 | 350 | MLS-MLG | III, VI |
| | | 24 | 30F70NC | 7.5 | 70.0 | 17.8 | 9.2 | 8.6 | 0.510 | 1.0 | 400 | SWG | III, VI |
| | | 25 | 20F80NC | 5.0 | 80.0 | 11.4 | 6.8 | 4.6 | 1.000 | 1.6 | 227 | SWG | III, VI |
| | | 26 | 100F | 25.0 | 0.0 | 45.5 | 20.1 | 25.4 | 0.006 | 0.6 | 10 | CI | III, VI |
| | | 27 | 100NC | 0.0 | 100.0 | – | – | – | 2.000 | 0.8 | 26 | SPG-GPS | III, VI |

Abbreviations:

Notes for testing conditions#:

I: 50 kPa, 0.06 mm/min; II: 100 kPa, 0.06 mm/min; III: 200 kPa, 0.06 mm/min; IV: 200 kPa, 0.6 mm/min; V: 200 kPa, 6.0 mm/min; VI: 400 kPa, 0.06 mm/min.

Notes for sample labels*:

K: Kaolin; GS: Glass sand; GB: Glass beads; GC: Coarse fraction made of 50% glass sand and 50% glass beads; F: Fine fraction made of 25% kaolin and 75% silt; NC: Coarse fraction made of 50% river sand and 50% crushed granite gravels.

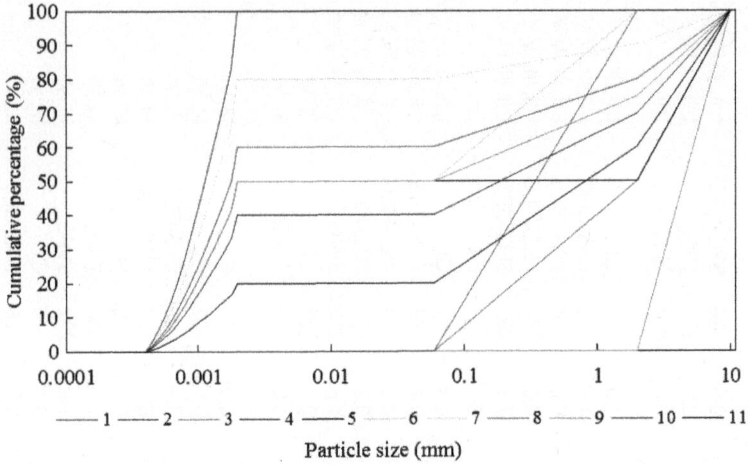


*Figure 4.1* Particle size distribution of the samples in the first category with single/binary/ternary fraction/s of kaolin, glass sand and glass beads.

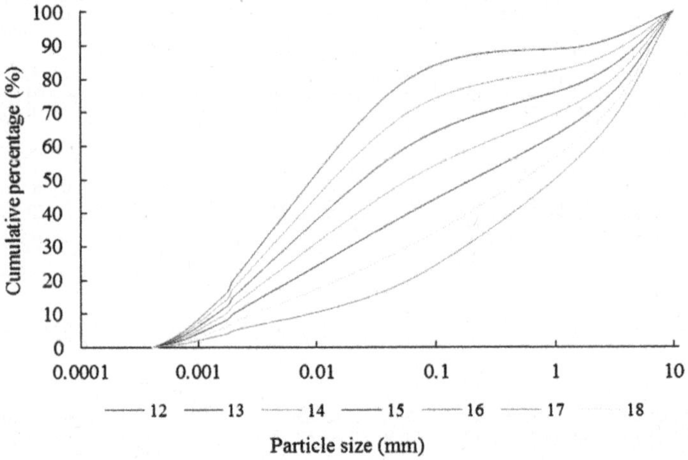


*Figure 4.2* Particle size distribution of the samples in the second category containing kaolin, silt, glass sand and glass beads.

For convenience, the abbreviations, KG is used to denote the kaolin and coarse fraction (glass sand and glass beads) mixture, and FG for mixture of kaolin, silt and coarse fraction, and FN for kaolin, silt, river sand and crushed gravels.

Classification of the samples by means of the Casagrande's plasticity chart is shown in Figure 4.4. It can be easily observed that there is a positive linear relationship between liquid limit and plasticity index for all the samples. Sample 1 is classified as CH with its high plasticity index of 28.3% and liquid limit of 50.3%. Samples 5, 12, 17, 19, 20 and 26 are plotted in the CI area owing to their moderate plasticity indices and

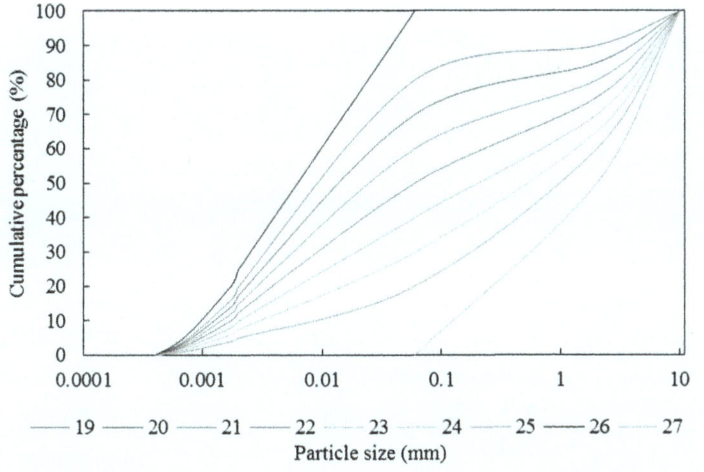


*Figure 4.3* Particle size distribution of the samples in the third category composed of kaolin, silt, natural sand and gravel.

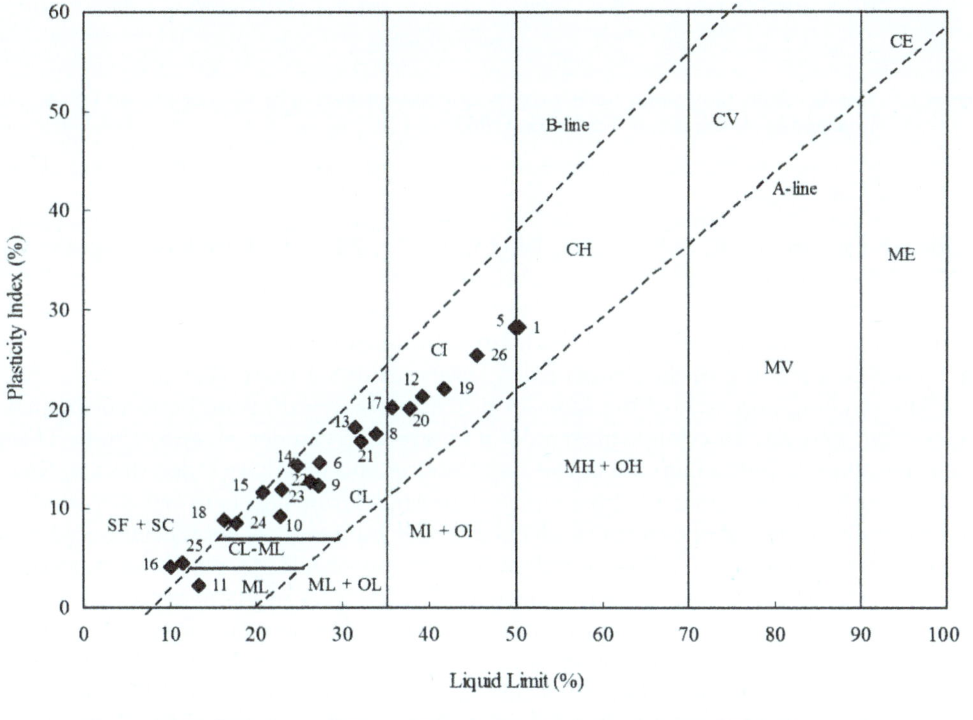


*Figure 4.4* Distribution of the samples on the Casagrande's (1948) plasticity chart.

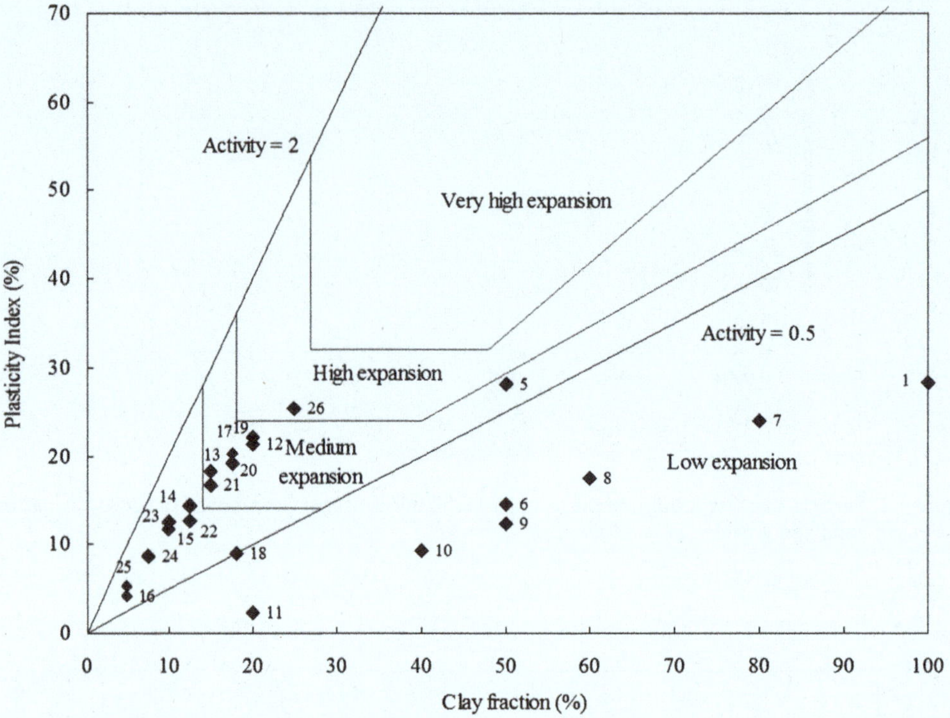


*Figure 4.5* Testing results plotted on activity charts originally proposed by van der Merwe (1964) and modified by Williams and Donaldson (1980).

liquid limits. Samples 6, 8, 9, 10, 13, 14, 15, 21, 22, 23 and 24 are located in the CL area, while Sample 11 is plotted in the ML area for its lower plasticity index of 2.3% and liquid limit of 13.4%. The remaining Samples 16, 18 and 25 fall into the SF + SC area with their low liquid limits. The classification symbols for the samples are listed in Table 4.2 according to the British Soil Classification Standard (BSCS).

The swelling potential of the samples based on the relationship between the plasticity index and clay fraction is presented in Figure 4.5 (van der Merwe, 1964). There is clearly a nearly linear increase in the clay fraction and plasticity index of samples of KG mixtures. There is also an approximately linear relationship between clay fraction and plasticity index of the samples of FG and FN mixtures. It should be noted that determination of plasticity limit involves significant variability which may account for the observed departures from the linearity. The samples of FG mixtures with glass sand and glass beads as the coarse fraction have relatively lower plasticity indices when compared with the corresponding samples of FN mixtures, e.g. Sample 12 compared to 19, though they have the same clay fraction. Sample 5 stands outside the main trend as an exception for the reason that only the kaolin fraction is tested for determining the plasticity index and the other component (glass beads) has no influence on the value of $I_P$.

### 4.2.2 Test equipment, sample preparation and testing procedures

#### *Test equipment*

Four strain-controlled digital direct/residual shear boxes (Fig. 3.3) were used in this study. The shear box consisted of a micro processor control unit, a large on-board LCD screen display and measuring devices. It was equipped with two digital dial gauges (range of 0 to 15 mm with a resolution of 0.002 mm), measuring vertical displacements and horizontal shear displacements. Normal stress was applied to the specimen through a lever loading device. Shear force was measured by means of two proving rings equipped with an electronic digital dial gauge inside, readable to 2 microns. These proving rings had measuring capacity of 2.0 kN and 4.5 kN, giving a probable accuracy of ±0.2 kPa and ±0.4 kPa, respectively. The former was used to measure shear force under low normal stress, while the latter was necessary when normal stresses and/or rates of shear displacement were high. This apparatus accepts square specimens of 100 mm × 100 mm × 40 mm. The shearing rate can be controlled to vary from $10^{-5}$ to 10 mm/min.

Horizontal and vertical displacements and shear force were recorded using a personal computer equipped with a data logger whose sampling frequency can be set to 1 record per 0.002 mm of shear displacement.

Calibration tests were conducted by using glass ballotini with a narrow size range of 0.2 to 0.3 mm under normal effective stresses of 10, 50 and 100 kPa, showing a linear ultimate strength envelope, passing through the origin; this indicates that errors associated with frictional resistance in the equipment are negligible (Lupini *et al.*, 1981). The gap between the shear boxes which restrain the specimen could be adjusted to ensure that there was no metal-to-metal sliding contact. Some difficulty was experienced in testing uniform glass sand and natural sand as particles could become jammed in the gap. The low rate of 0.06 mm/min could be achieved as soon as the shear started. It was, however, observed that the motor drivers of the shear boxes needed about 3 minutes to accelerate to the desired rates of shear displacement of 0.6 and 6.0 mm/min.

#### *Sample preparation*

All constituent materials were oven dried at a low temperature of 105 ± 0.5°C to avoid the influence of high temperature on the structure of kaolin. Each dry constituent material was thoroughly mixed, heaped and divided by the quartering method (dividing a circular heap, by diameters at right angles, into four equal parts, removing two diagonally opposite quarters, and thoroughly mixing the two remaining quarters together) so as to obtain a representative mass. The quartering process was repeated over all constituent materials to get the necessary quantity of material for each sample in Table 4.2 to avoid having to re-use any part of it.

The sample was mixed by taking its constituent/s. The mixture was then placed into a plastic bag and shaken until it appeared homogeneous. Each prepared sample was divided into several duplicate specimens by again using the quartering method for the series of tests under different normal stresses and at different shearing rates as listed in Table 4.2. Each sample of KG mixture was divided into 5 specimens, with an exception for Samples 2 and 3 that had 6 specimens. Each sample of FG mixture was divided into 5 specimens, except Sample 12 with 6 specimens and Samples 17 and 18 with 2 specimens each. Each sample of FN mixture was divided into 2 specimens. Each specimen was to be tested under one of the six testing conditions which differ from each other as to the magnitude of normal stress and the shearing rate (Tables 4.2 and 4.3).

### *Testing procedures*

All tests were set up by following the same compaction procedure. Each test specimen was prepared in the shear box in three layers of equal thickness. Each layer was compacted by means of a tamping rod (with a square end of 10 mm × 10 mm), subjected to 30 blows of a 2.5 kg hand rammer falling through 450 mm. Since specimens of each sample (e.g. Specimens 1I, 1II, etc. from Sample 1) had identical mass and were compacted in the same volume, they had identical dry densities and void ratios before saturation and therefore the same water contents after fully saturated. The specimen then was subjected to approximately 24-hour saturation, and then an about 24-hour consolidation period under the pre-specified vertical stress. Vertical deformation of the specimen was recorded during consolidation. After vertical deformation ceased, the upper and lower shear boxes were separated for approximately 0.2 mm and shearing was conducted at a constant shearing rate as specified in Table 4.2.

The testing procedures also involved the following:

1. Untested specimens were used for each test;
2. Each specimen was sheared until the relative horizontal displacement of approximately 14 mm was achieved;
3. A small seating pressure of 4.21 kPa was applied on the load cap during saturation;
4. Initial reading of vertical displacement was reset to zero at the beginning of each test stage;
5. Test data were recorded at a frequency of one record per 2 μm horizontal displacement;
6. The contacting surfaces of the shear box were lubricated with grease; and
7. The laboratory was maintained at the temperature of 19.5 ± 0.5°C.

## 4.3 RESULTS AND DISCUSSION

The assumption that the normal and shear stresses are distributed uniformly across the shear plane is adopted as the basis for calculation of the normal stress ($\sigma$) and shear stress ($\tau$) by dividing the measured vertical load and shear force by their corresponding areas. Since the area in contact with normal load remains constant as the sectional area of sample chamber (in this study, 100 mm × 100 mm) during testing, no area corrections were made for normal stress. However, the soil-to-soil area in the shear plane decreases as the test progresses, area corrections were applied only to the shear stress. The corrected area ($A_{CS}$) in the shear plane for tests conducted with a square box, as the one used in this study, is calculated from $A_{CS} = A_I - (W \times \delta_h)$, where $A_I$ is the initial area, $W$ is the width of the soil specimen, and $\delta_h$ is the horizontal displacement (Kim *et al.*, 2006; Xu *et al.*, 2007; Bareither *et al.*, 2008).

*Table 4.3* Testing results of composite soils.

| *Testing conditions* | | | | *Shear stage* | | |
|---|---|---|---|---|---|---|
| *Sample no.* | *Test no.* | *Normal effective stress σ′ (kPa)* | *Shearing rate (mm/min)* | *Horizontal displacement @Res. (mm)* | *Vertical displacement @Res. (mm)* | *Residual stress ratio $\tau_R/\sigma'$* |
| 1 | 1I | 50 | 0.06 | 3.56 | −0.10 | 0.33 |
| | 1II | 100 | 0.06 | 5.15 | −0.26 | 0.32 |
| | 1III | 200 | 0.06 | 10.63 | −0.49 | 0.31 |
| | 1IV | 200 | 0.60 | 11.60 | −0.34 | 0.45 |
| | 1V | 200 | 6.00 | 13.40 | −0.16 | 0.48 |
| 2 | 2I | 50 | 0.06 | 5.02 | −0.17 | 0.67 |
| | 2II | 100 | 0.06 | 6.07 | 0.17 | 0.70 |
| | 2III | 200 | 0.06 | 7.14 | 0.32 | 0.49 |
| | 2IV | 200 | 0.60 | 9.33 | −0.27 | 0.39 |
| | 2V | 200 | 6.00 | 11.62 | −0.10 | 0.37 |
| | 2VI | 400 | 0.06 | 11.07 | −0.24 | 0.37 |
| 3 | 3I | 50 | 0.06 | 7.48 | 1.23 | 0.75 |
| | 3II | 100 | 0.06 | 6.14 | 0.47 | 0.72 |
| | 3III | 200 | 0.06 | 7.05 | 0.51 | 0.61 |
| | 3IV | 200 | 0.60 | 4.29 | 0.38 | 0.39 |
| | 3V | 200 | 6.00 | 7.30 | 0.38 | 0.45 |
| | 3VI | 400 | 0.06 | 10.92 | 1.12 | 0.39 |
| 4 | 4I | 50 | 0.06 | 4.39 | 0.20 | 0.73 |
| | 4II | 100 | 0.06 | 3.34 | 0.06 | 0.70 |
| | 4III | 200 | 0.06 | 7.89 | −0.25 | 0.59 |
| | 4IV | 200 | 0.60 | 6.34 | −0.30 | 0.42 |
| | 4V | 200 | 6.00 | 6.16 | −0.04 | 0.45 |
| | 4VI | | | | | 0.49 |
| 5 | 5I | 50 | 0.06 | 6.10 | −0.16 | 0.67 |
| | 5II | 100 | 0.06 | 5.92 | −0.17 | 0.67 |
| | 5III | 200 | 0.06 | 7.13 | −0.29 | 0.55 |
| | 5IV | 200 | 0.60 | 7.27 | −0.17 | 0.45 |
| | 5V | 200 | 6.00 | 10.17 | −0.16 | 0.36 |
| 6 | 6I | 50 | 0.06 | 7.44 | −0.36 | 0.64 |
| | 6II | 100 | 0.06 | 5.45 | −0.16 | 0.65 |
| | 6III | 200 | 0.06 | 5.67 | −0.19 | 0.46 |
| | 6IV | 200 | 0.60 | 9.47 | −0.25 | 0.39 |
| | 6V | 200 | 6.00 | 8.57 | −0.10 | 0.44 |
| 7 | 7I | 50 | 0.06 | 6.69 | −0.21 | 0.59 |
| | 7II | 100 | 0.06 | 8.26 | −0.33 | 0.50 |
| | 7III | 200 | 0.06 | 8.98 | −0.44 | 0.45 |
| | 7IV | 200 | 0.60 | 7.32 | −0.26 | 0.38 |
| | 7V | 200 | 6.00 | 11.53 | −0.21 | 0.42 |
| 8 | 8I | 50 | 0.06 | 8.85 | −0.17 | 0.62 |
| | 8II | 100 | 0.06 | 6.70 | −0.63 | 0.54 |
| | 8III | 200 | 0.06 | 8.61 | −0.34 | 0.48 |
| | 8IV | 200 | 0.60 | 9.08 | −0.23 | 0.44 |
| | 8V | 200 | 6.00 | 8.89 | −0.06 | 0.40 |
| 9 | 9I | 50 | 0.06 | 8.67 | −0.12 | 0.62 |
| | 9II | 100 | 0.06 | 7.90 | −0.20 | 0.59 |
| | 9III | 200 | 0.06 | 9.75 | −0.18 | 0.49 |
| | 9IV | 200 | 0.60 | 6.74 | −0.19 | 0.30 |
| | 9V | 200 | 6.00 | 9.43 | −0.06 | 0.42 |

(continued)

*Table 4.3* Continued.

| *Testing conditions* | | | | *Shear stage* | | |
|---|---|---|---|---|---|---|
| *Sample no.* | *Test no.* | *Normal effective stress $\sigma'$ (kPa)* | *Shearing rate (mm/min)* | *Horizontal displacement @Res. (mm)* | *Vertical displacement @Res. (mm)* | *Residual stress ratio $\tau_R/\sigma'$* |
| 10 | 10I | 50 | 0.06 | 6.93 | −0.01 | 0.65 |
| | 10II | 100 | 0.06 | 8.06 | −0.17 | 0.60 |
| | 10III | 200 | 0.06 | 7.90 | −0.25 | 0.52 |
| | 10IV | 200 | 0.60 | 8.33 | −0.24 | 0.46 |
| | 10V | 200 | 6.00 | 12.57 | −0.15 | 0.38 |
| 11 | 11I | 50 | 0.06 | 9.49 | 0.33 | 0.66 |
| | 11II | 100 | 0.06 | 9.31 | 0.15 | 0.62 |
| | 11III | 200 | 0.06 | 8.48 | 0.08 | 0.58 |
| | 11IV | 200 | 0.60 | 7.57 | −0.39 | 0.61 |
| | 11V | 200 | 6.00 | 12.70 | 0.04 | 0.61 |
| 12 | 12I | 50 | 0.06 | 10.16 | −0.58 | 0.62 |
| | 12II | 100 | 0.06 | 8.15 | −0.56 | 0.55 |
| | 12III | 200 | 0.06 | 11.17 | −0.58 | 0.48 |
| | 12IV | 200 | 0.60 | 11.12 | −0.35 | 0.45 |
| | 12V | 200 | 6.00 | 11.59 | −0.35 | 0.46 |
| | 12VI | 400 | 0.06 | 12.23 | −0.49 | 0.44 |
| 13 | 13I | 50 | 0.06 | 8.62 | 0.16 | 0.65 |
| | 13II | 100 | 0.06 | 6.94 | −0.17 | 0.58 |
| | 13III | 200 | 0.06 | 9.32 | −0.04 | 0.50 |
| | 13IV | 200 | 0.60 | 11.33 | −0.20 | 0.40 |
| | 13VI | 400 | 0.06 | 10.62 | −0.42 | 0.45 |
| 14 | 14I | 50 | 0.06 | 9.30 | 0.13 | 0.69 |
| | 14II | 100 | 0.06 | 9.24 | −0.03 | 0.61 |
| | 14III | 200 | 0.06 | 7.21 | −0.09 | 0.49 |
| | 14IV | 200 | 0.60 | 12.97 | −0.15 | 0.50 |
| | 14VI | 400 | 0.06 | 10.12 | −0.37 | 0.46 |
| 15 | 15I | 50 | 0.06 | 8.16 | 0.11 | 0.68 |
| | 15II | 100 | 0.06 | 8.38 | −0.13 | 0.65 |
| | 15III | 200 | 0.06 | 8.20 | −0.16 | 0.50 |
| | 15IV | 200 | 0.60 | 13.41 | −0.12 | 0.53 |
| | 15VI | 400 | 0.06 | 10.73 | −0.16 | 0.46 |
| 16 | 16I | 50 | 0.06 | 8.82 | 0.42 | 0.70 |
| | 16II | 100 | 0.06 | 10.89 | 0.32 | 0.70 |
| | 16III | 200 | 0.06 | 10.91 | 0.31 | 0.55 |
| | 16IV | 200 | 0.60 | 12.96 | 0.17 | 0.56 |
| | 16VI | 400 | 0.06 | 7.63 | −0.47 | 0.48 |
| 17 | 17III | 200 | 0.06 | 9.94 | −0.03 | 0.49 |
| | 17VI | 400 | 0.06 | 10.60 | −0.59 | 0.45 |
| 18 | 18III | 200 | 0.06 | 8.32 | −0.13 | 0.52 |
| | 18VI | 400 | 0.06 | 8.72 | −0.38 | 0.47 |
| 19 | 19III | 200 | 0.06 | 8.13 | −0.47 | 0.56 |
| | 19VI | 400 | 0.06 | 19.50 | −0.54 | 0.41 |
| 20 | 20III | 200 | 0.06 | 8.37 | −0.36 | 0.60 |
| | 20VI | 400 | 0.06 | 9.55 | −0.51 | 0.52 |
| 21 | 21III | 200 | 0.06 | 7.85 | −0.12 | 0.64 |
| | 21VI | 400 | 0.06 | 18.04 | −0.59 | 0.62 |

*(continued)*

*Table 4.3* Continued.

| *Testing conditions* | | | | *Shear stage* | | |
|---|---|---|---|---|---|---|
| *Sample no.* | *Test no.* | *Normal effective stress $\sigma'$ (kPa)* | *Shearing rate (mm/min)* | *Horizontal displacement @Res. (mm)* | *Vertical displacement @Res. (mm)* | *Residual stress ratio $\tau_R/\sigma'$* |
| 22 | 22III | 200 | 0.06 | 9.01 | −0.07 | 0.67 |
| | 22VI | 400 | 0.06 | 23.13 | −0.72 | 0.63 |
| 23 | 23III | 200 | 0.06 | 8.80 | 0.03 | 0.69 |
| | 23VI | 400 | 0.06 | 17.47 | −0.68 | 0.65 |
| 24 | 24III | 200 | 0.06 | 8.97 | −0.35 | 0.72 |
| | 24VI | 400 | 0.06 | 10.27 | −0.77 | 0.72 |
| 25 | 25III | 200 | 0.06 | 9.09 | 0.90 | 0.94 |
| | 25VI | 400 | 0.06 | 13.79 | −0.32 | 0.76 |
| 26 | 26I | | | | | 0.57 |
| | 26II | | | | | 0.52 |
| | 26III | 200 | 0.06 | 7.95 | −0.53 | 0.44 |
| | 26IV | | | | | 0.43 |
| | 26VI | 400 | 0.06 | 16.55 | −0.45 | 0.38 |
| 27 | 27III | 200 | 0.06 | 12.03 | 1.47 | 0.97 |
| | 27VI | 400 | 0.06 | 14.39 | 0.14 | 0.89 |

Abbreviation: @Res: at residual state.

In the following expatiation on the testing results, residual strength $\tau_R$ was used as the shear stress operating after failure to the residual state. The term stress ratio $\tau/\sigma'$, where $\sigma'$ is the effective normal stress operating on the failure surface, when at residual state, was defined as the residual friction coefficient or residual stress ratio. The residual friction angle $\varphi'_R$ was derived from the tangent of this stress ratio using $\varphi'_R = \arctan(\tau_R/\sigma')$. The terms "clay fraction", "fine fraction" and "coarse fraction" were used for the percentages by weight of particles smaller than 2 microns, smaller than 63 microns (clay and silt) and greater than 63 microns in diameter, respectively. The nature of the shear surface formed in the tests was ascertained primarily by attempting to split the moist specimens on the potential shear surface after removal from the apparatus and visual examination.

The results of tests are summarized in Table 4.3, where they are identified by the test numbers indicating the specimen type and testing conditions. This systematically tabulates the residual strength and corresponding shear displacement and the change in sample height at this displacement. The stress ratio ($\tau/\sigma'$) curves are plotted as a function of shear displacement in Figure 4.6(a–e).

### 4.3.1 Compressibility

Perhaps originating from the early work on the '$e - \log p'$' relation proposed by Terzaghi, void ratio $e$ has been chosen as one of the most important state variables to characterize the behavior of soils. This has partly influenced the work of Roscoe and co-workers (Roscoe *et al.*, 1958, 1963; Roscoe & Burland, 1968; Schofield & Wroth,

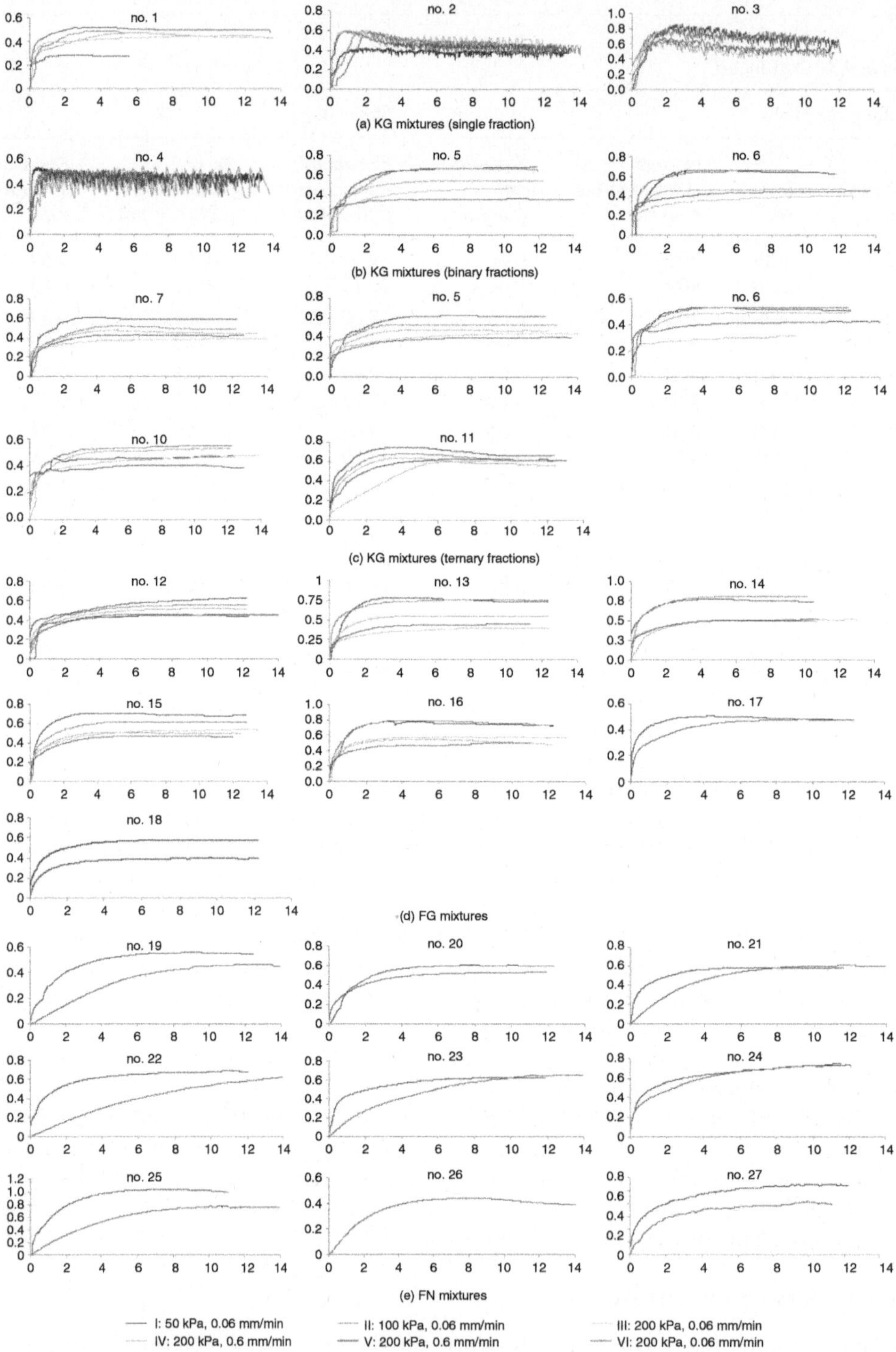


*Figure 4.6(a–e)* Stress ratio, $\tau/\sigma'$, vs. shear displacement (mm) plots for intermediate direct shear tests. The abscissa represents shear displacement in mm and the ordinate represents stress ratio.

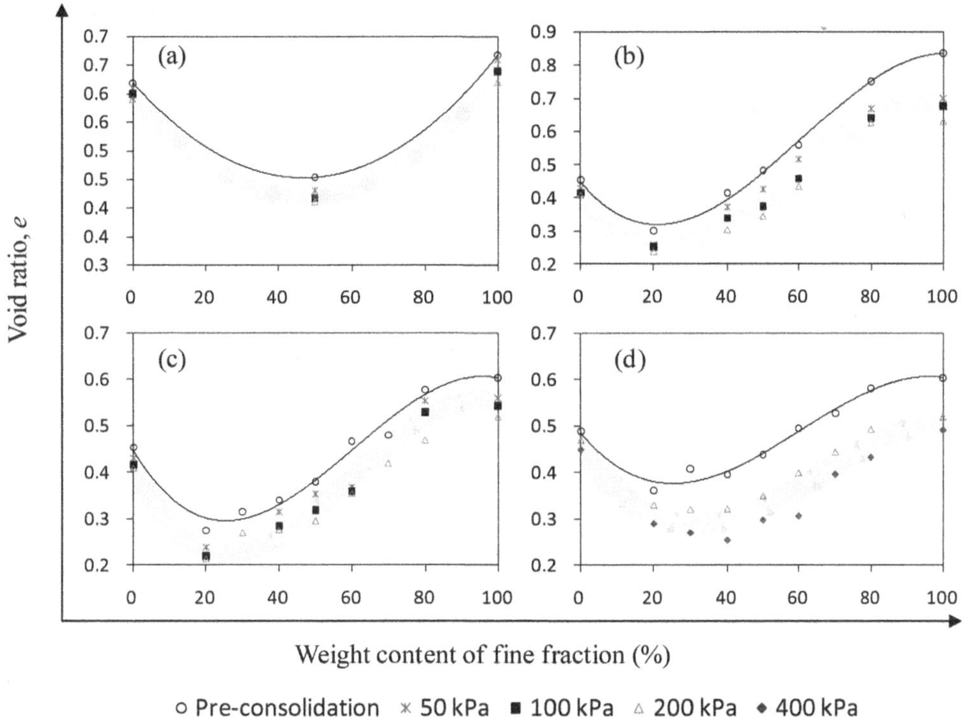


*Figure 4.7* Void ratio appears to be a function of gradation of soil mixtures: (a) granular mixtures; (b) KG mixtures; (c) FG mixtures; and (d) FN mixtures.

1968; Schofield, 2005) in their choice of ($p'$, $q$, $e$) as the state variables in critical state soil mechanics (where $p' =$ mean effective stress and $q =$ deviatoric stress) and the formulation of the so-called steady-state concept (Poulos, 1981).

Both the pre-consolidation and post-consolidation void ratios of each specimen were calculated according to the sample preparation and measurement of vertical displacement due to consolidation. The void ratios, as functions of the percentage by weight of the fine fraction, are presented in Figure 4.7(a–d). Each figure represents the results of a set of specimens from the same type of mixture. Figure 4.7a indicates changes in void ratio with increasing weight content of glass sand (0, 50, 100%) for mixtures with glass sand and glass beads. In Figure 4.7b, experimental points show an initial reduction with increasing fine fraction (kaolin) and decreasing coarse fraction (glass sand + glass beads), reach a minimum value of approximately 0.25 corresponding to the fine fraction between 20% and 30%, and increasing all the way up to the value obtained for the fine fraction alone. Figure 4.7c shows the changing pattern of void ratio with increasing of the fine fraction (kaolin + silt) and decreasing of the coarse fraction (glass sand + glass beads). For the mixtures with kaolin, silt, natural sand and natural gravel (Fig. 4.7d), the same pattern is observed.

In brief, the void ratio of the mixture, after consolidation, depends on the fractional concentration of each particle size population. Increase in the fine fraction from zero until a threshold, 20–30% in this case, causes a decrease in the void ratio of the mixture. Beyond the threshold, a further increase in the fine fraction causes an increase in the void ratio up to the values obtained for the fine material alone. The void ratio curves for the same family of mixtures, when consolidated under different normal stresses, follow a similar trend. Furthermore, the void ratio of the mixture decreases as the level of imposed normal stress increases.

Theoretical relationships can be used to express the changing of void ratio, when binary mixtures are considered, using ideal packing states of floating and non-floating oversized particles within a fine matrix (Fragaszy *et al.*, 1990; Vallejo & Mawby, 2000; Simoni & Houlsby, 2006). When a small amount of fines (e.g. kaolin in our case) are added into the host coarse material (e.g. glass sand and glass beads), the non-floating coarse particles are in contact and the bulk volume of the clay (clay volume and their enclosed void space) tend to occupy a portion of the voids built up by coarse skeleton. If the fines are continued to be added, there must be a threshold at which the void of coarse skeleton is to be completely occupied by the bulk volume of the fines. At this point, the mixture owns the minimum void ratio (McGeary, 1961; Rogers *et al.*, 1994; Vallejo & Mawby, 2000; Li, 2013). Beyond this threshold, if more fines are added, they will play as a separator of the coarse particles and make the coarse particles float or scatter within the fine matrix. According to this, the achievable void ratio of composite soil is strongly dependent upon the fractional concentration of each particle size population, assuming the same compaction effort is used.

An obvious difference can be observed when plotting the void ratio trends of KG mixtures and that of FG mixtures in Figure 4.8. Referring to Table 4.2, KG mixtures use only kaolin as the fine fraction, whereas FG mixtures use kaolin and silt (weight ratio of 1:3) as their fine fraction. The coarse fractions (glass sand + glass beads) are the same in these two mixtures. The relatively higher void ratios of KG mixtures may result from the poor gradation by missing silt-size particles. As shown in Figure 4.9, it is also evident that the mixture with natural sand and gravel as the coarse fraction has higher compressibility when compared to that with glass sand and glass beads. Clearly angularity hinders particle mobility and its ability to gain denser packing configurations. In the extreme case of low sphericity, platy particles bridge gaps over particles, thus creating large open voids (Guimaraes, 2002). This observation is in good agreement with those of previous researchers (Shakoor & Cook, 1990; Fragaszy *et al.*, 1990; Vallejo & Mawby, 2000; Vallejo, 2001; Simoni & Houlsby, 2006; Cho *et al.*, 2006).

In Figure 4.8, the difference in void ratio, as highlighted using dashed lines appears to increase with the fine fraction, which indicates that after threshold, the void ratios of the mixtures are dominated by the content and properties (here, gradation) of fine fraction if the coarse fractions are the same. On the contrary, in Figure 4.9, the difference in void ratio decreases with the fine fraction, indicating the void ratio is dominated by the coarse fraction properties (here, sphericity and surface roughness) if the fine fractions are the same. Previous researchers have indicated that the shape of coarse particles affects attainable void ratios in soils with fines content and the maximum and minimum void ratios increase with decrease in roundness or sphericity (Shimobe & Moroto 1995; Miura *et al.*, 1998; Cubrinovski & Ishihara, 2002; Santamarina & Cho, 2004). The irregular particles produce bridges between themselves causing a high local porosity underneath.

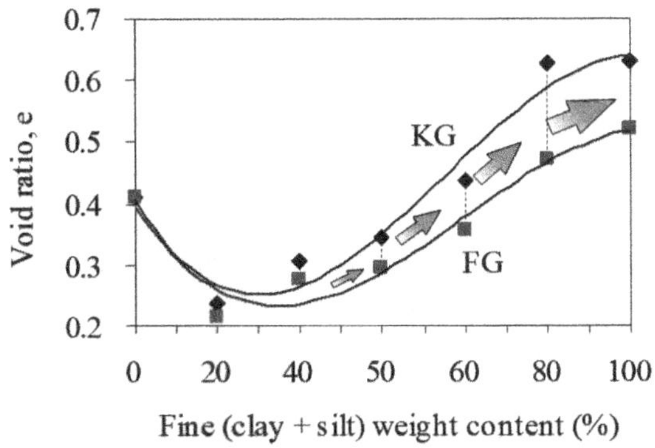


*Figure 4.8* Comparison of void ratios between KG and FG mixtures. The two series have the same coarse fraction. Only samples subjected to 200 kPa consolidation are taken for the clarity of the figure.

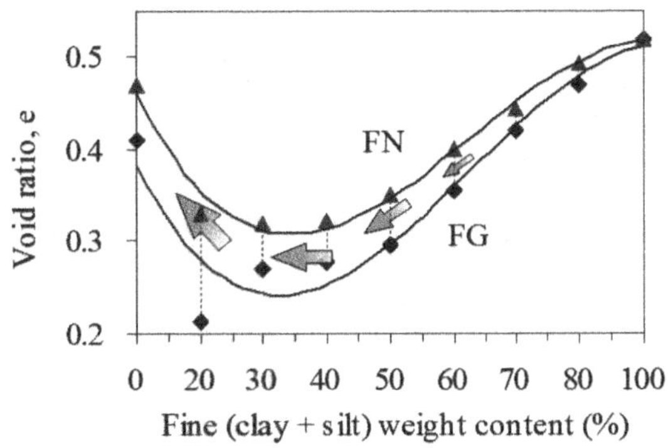


*Figure 4.9* Comparison of void ratios between FG and FN mixtures. These two series have the same fine fraction. Only mixtures subjected to 200 kPa consolidation are taken for the clarity of the figure.

### 4.3.2 Water content after shearing

The water content of each sheared specimen was determined immediately after removal out from the apparatus and wiping off the free water on the surfaces. Two pieces, each weighing approximately 40 g, were cut from the top and bottom third layers of each specimen. A strip of approximately 10 mm thick and 60 mm wide was cut from the middle of the slip zone. This strip was then divided into three parts and the average water content for the slip zone was determined. This enabled examination of the difference between water content of shear zone and that of other parts of sheared specimen. The water contents of Samples 2, 3, 4, and 27, which are composed of only coarse fraction, were not determined due to free drainage of these samples. Since water contents measured from the top and bottom layers were approximately the same for each specimen with an overall average standard deviation of 0.05%, their average was used for comparison with the average of the shear zone as presented in Figure 4.10.

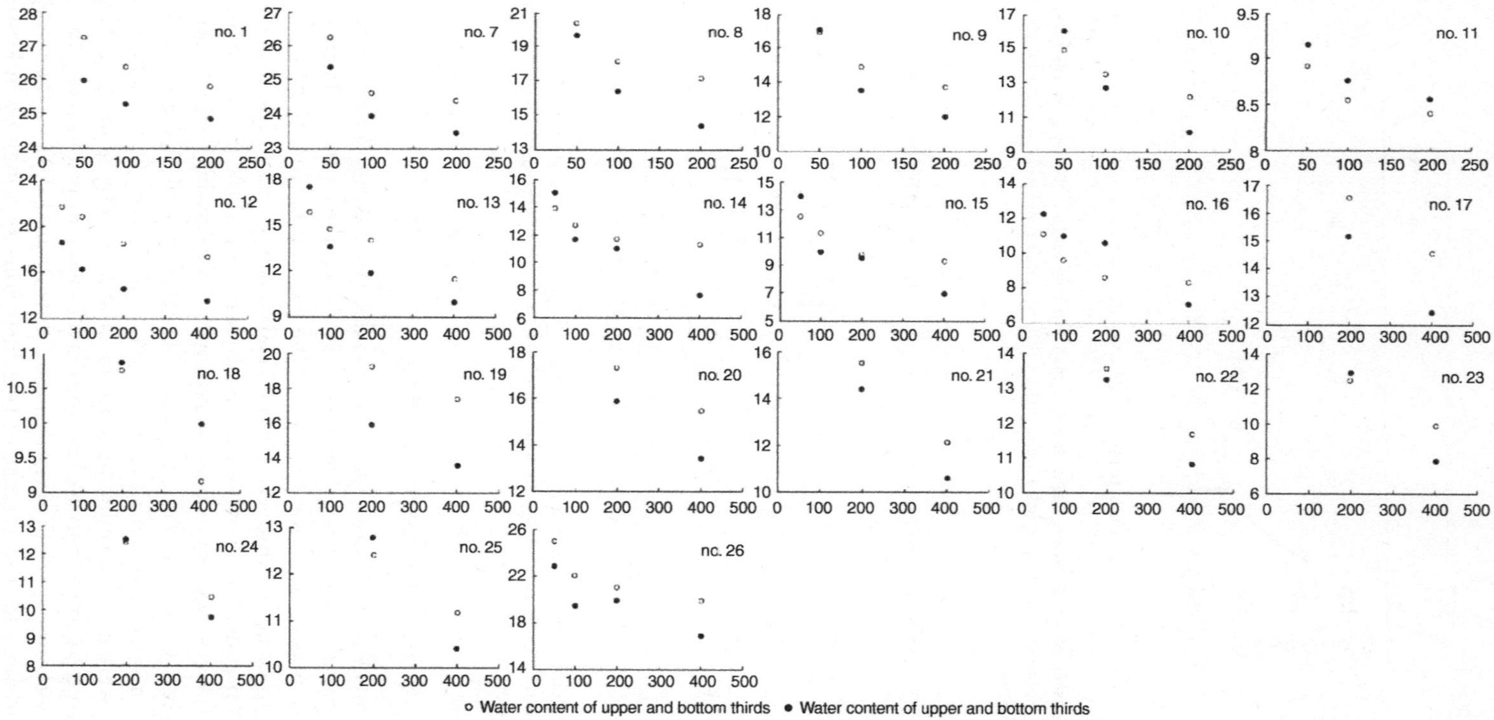


*Figure 4.10* Comparison of water contents of the shear zone with other parts of the sheared specimens. The abscissa is effective normal stress in kPa and the ordinate is water content in percent.

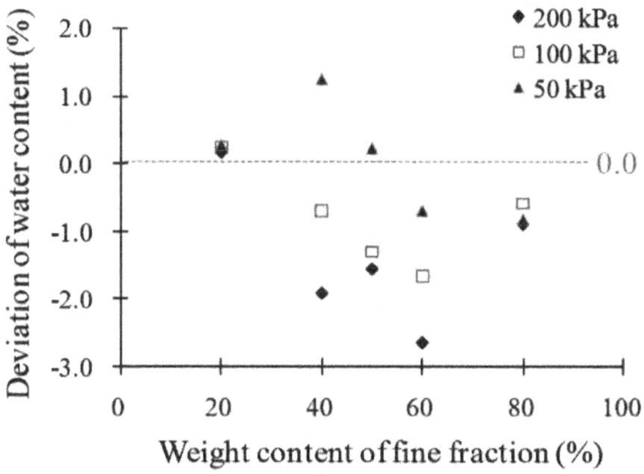


*Figure 4.11* Differences in water content of the shear zone and the surrounding layers of the sheared specimens (KG mixtures).

In Samples 1, 7, and 8 of KG mixtures, water content of the shear zone is lower than that of the surrounding layers. For Samples 9 and 10, water content in the shear zone is slightly higher than other parts when sheared under 50 kPa (Test/Specimen 9I) but lower when sheared under 100 kPa (9II) and 200 kPa (9III). In the case of Sample 11, water content of the shear zone is consistently higher than that of other parts regardless of normal stress.

The differences between water contents of the shear zone and other parts appear to be strongly dependent upon both the gradation and magnitude of normal stress as shown in Figure 4.11. It decreases gradually from positive to negative with the increase in weight content of fine fraction of samples. In other words, high proportion of fine fraction causes lower water content in the shear zone; low proportion results in higher one, compared with other parts. This indicates that the high proportion of fine fraction makes the shear zone of specimen more susceptible to contracting during shearing and vice versa. In addition, it is also found that for each sample, with the increase of normal stress, the difference in water content of shear zone and other parts decreases. For example, Sample 9 (50% kaolin) has greater water content of shear zone than other parts when sheared under 50 kPa (9I). However, under 100 (9II) and 200 kPa (9III), it has less water content of shear zone than outer zones. This may indicate that high normal stress would suppress the dilation of shear zone during shearing. This phenomenon will be discussed in Section 4.3.3 in terms of the structure of shear zone.

For the samples of FG mixtures, differences in water content of the shear zone and outer layers show the same trend as KG mixtures (Fig. 4.12). High proportion of fine fraction causes higher water content in the shear zone; low proportion results in the lower one. With the intermediate proportion of fine fraction, higher water content in the shear zone is related to low normal stress and vice versa. At 50 kPa, the positive difference between water contents emerges at 60% fine fraction for Sample 13 of FG mixture as shown in Figure 4.12 and at 50% for Sample 9 of KG mixture as shown

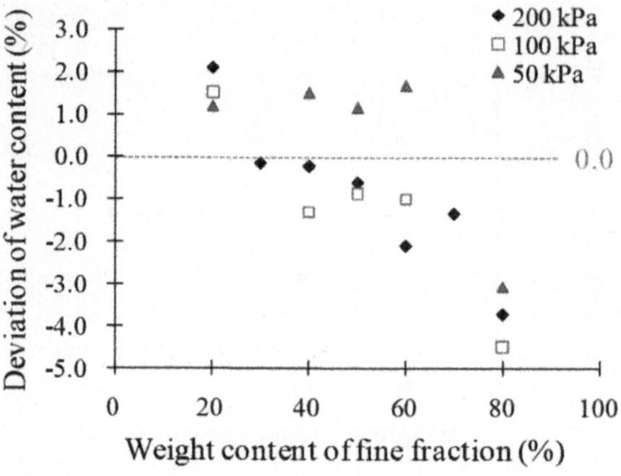


*Figure 4.12* Differences in water contents of shear zone and surrounding layers (FG mixtures).

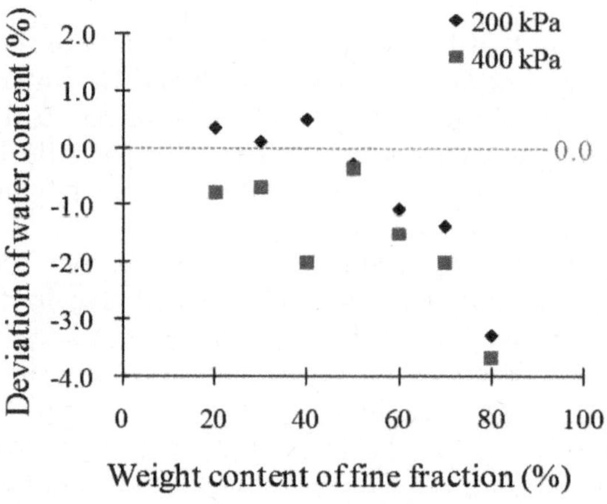


*Figure 4.13* Water content differences between shear zone and other parts (FN mixtures).

in Figure 4.11. The differences in the magnitudes and the percentages may be ascribed to the differences in these mixtures, particularly the silt content (Table 4.2).

For the samples of FN mixtures, water contents of the shear zone and other parts differ along the same trend as KG mixture (Fig. 4.13). Once again, the proportion of fine fraction determines the water content in the shear zone. Similarly, for the intermediate proportion of fine fraction, higher water content in the shear zone is attributable to low normal stress and vice versa. At 200 kPa, the positive difference in water content is first observed in Sample 23 (Fig. 4.13) which contains 40% fine fraction while this occurred in Sample 18 (80%) of FG mixture (Fig. 4.12). Substituting glass sand and glass beads by natural river sand and crushed gravels caused an increase in the relative magnitude of water content within the shear zones. It is appropriate to hypothesize that the increase can be attributed to angularity and roughness of coarse particles.

### 4.3.3 Granular materials

#### *Stress fluctuations*

The glass sand and beads, used to compose Samples 2, 3 and 4, are of uniform distribution, with the mean sizes of 0.32 mm and 4.2 mm, respectively. These materials were essentially free of particle breakage during shearing in the present study.

Regardless of the normal stresses and shearing rates, the stress–strain curves of Samples 2, 3 and 4 in Figure 4.6(a–e) show the standard pattern for dense sands: stresses decrease after the peak stress level and then flatten; vertical displacements decrease at the beginning, then increase and flatten. Note the pronounced fluctuations on the stress-strain curves. These fluctuations on the stress and vertical strain were widely observed by many researchers studying the shear behavior of granular soils (Skinner, 1969; Oda & Junichi, 1974; Shimizu, 1997; Fukuoka *et al.*, 2006) or composite soils (Pitman *et al.*, 1994; Patwardhan *et al.*, 1970; Clausen & Gabrielsen, 2002; Prakasha & Chandrasekaran, 2005; Li & Aydin, 2010), using various type of shear apparatuses including direct shear, simple shear, triaxial compression or ring shear. Shimizu (1997) indicated that the amplitude of the fluctuation is so large, especially for the dense sand, that it may cause the shear stress to momentarily drop down even to zero.

The fluctuation is considered to result from the interaction of particles that resist each other when they come into contact. This interaction may involve sliding and rotation (Oda & Kazama, 1998). For clarity, the results from Sample 2, tested at 50 kPa (Test/Specimen 2I), 100 kPa (2II) and 200 kPa (2III) and at the shearing rate of 0.06 mm/sec, are presented again in Figures 4.14 to 4.16, respectively. In each figure, the complete curves of stress ratio and vertical displacement versus horizontal displacement are presented in the upper plot, while portions of the curves are magnified in the lower plot. It is shown that the fluctuations can occupy a range of approximately 1/4 of the total value, causing an unacceptable error if not properly considered. Much effort was made to quantify the peak and bottom values of these fluctuations carefully, which were characterized by amplitude and wave length parameters in Figure 4.17(a–c).

#### *The Fast Fourier Transform (FFT) analysis*

The advanced data logging system attached to the direct shear apparatus enabled near analog sampling at a high frequency of one record per 0.002 mm horizontal displacement, resulting in a dataset of about 7000 readings for each test. In order to efficiently analyze this large dataset and especially to derive a deeper insight into the seemingly irregular fluctuations, the Fourier Transform (FT) was chosen as the mathematical technique for its suitability to deal with this type of data.

Fourier analysis is a family of mathematical techniques, named after Jean Baptiste Joseph Fourier (1768–1830), one of the most famous mathematicians. The essential concept of the Fourier analysis is that all signals can be decomposed into sinusoids. One form of Fourier analysis, Fourier Transform can be, in general, broken into four categories, Fourier Transform (FT), Fourier Series (FS), Discrete Time Fourier Transform

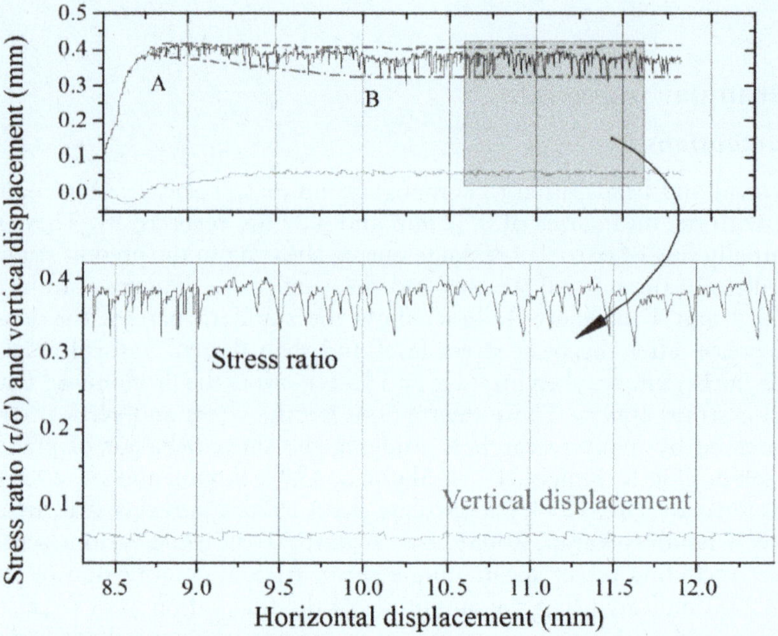


*Figure 4.14* Stress ratio ($\tau/\sigma'$) and corresponding vertical displacement versus horizontal displacement for Sample 2 tested at 50 kPa (Specimen 2I).

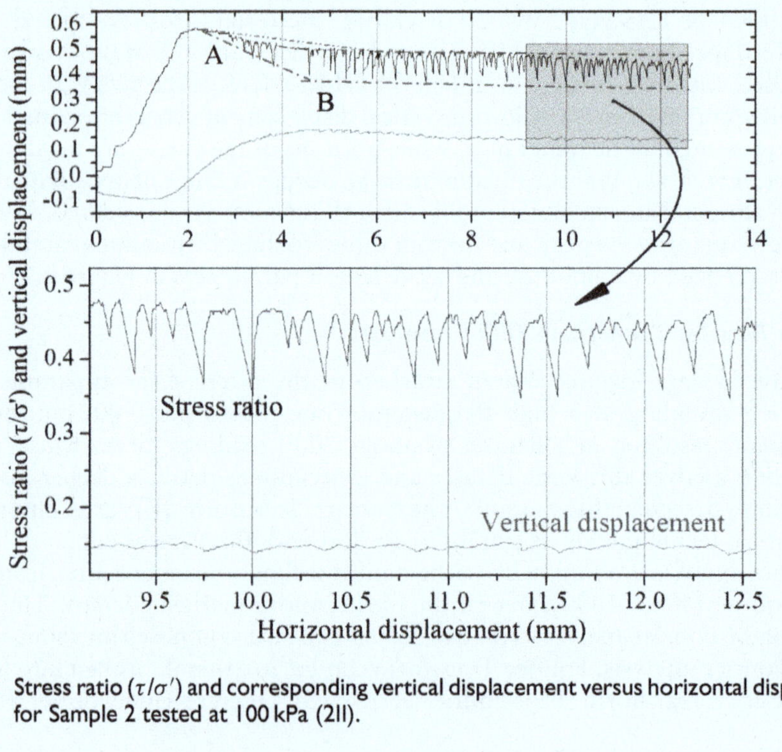


*Figure 4.15* Stress ratio ($\tau/\sigma'$) and corresponding vertical displacement versus horizontal displacement for Sample 2 tested at 100 kPa (2II).

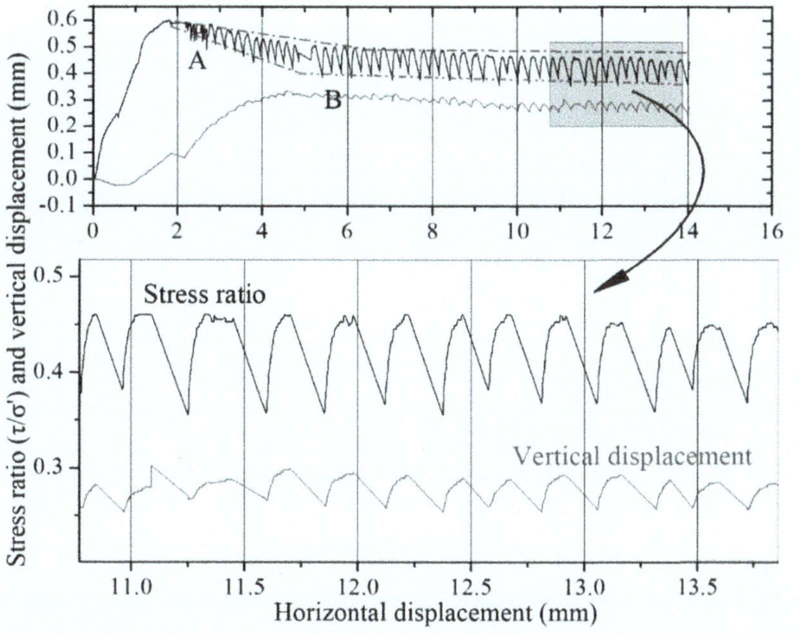


*Figure 4.16* Stress ratio ($\tau/\sigma'$) and corresponding vertical displacement versus horizontal displacement for Sample 2 tested at 200 kPa (2III).

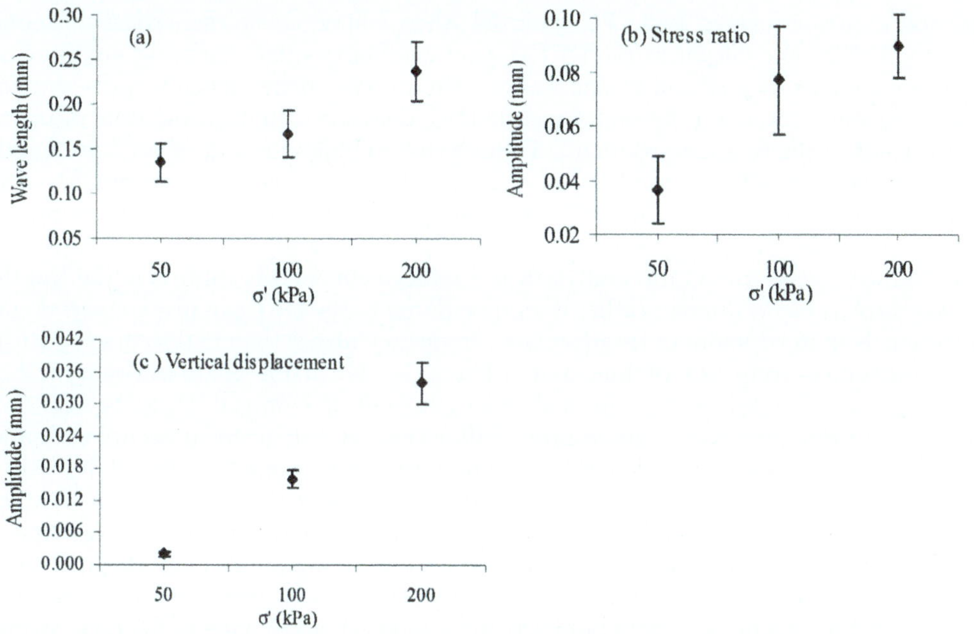


*Figure 4.17(a–c)* Fluctuation characteristics of stress ratio at residual state as a function of effective normal stress.

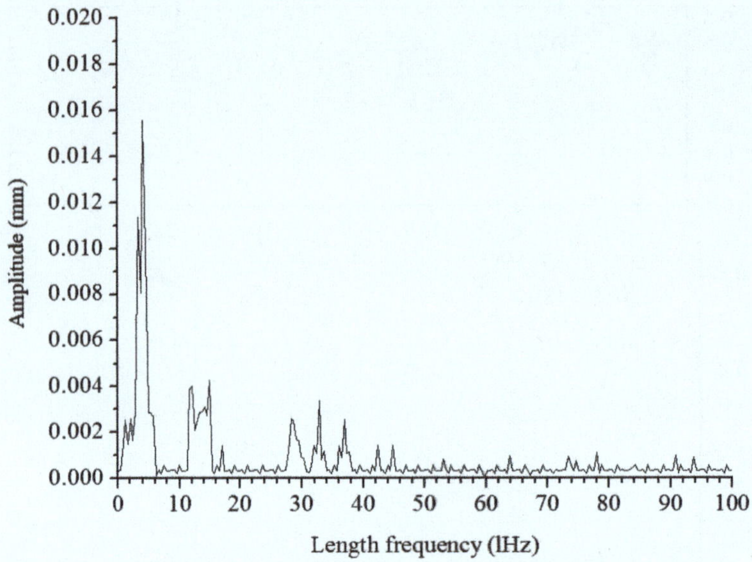


*Figure 4.18* The length frequency spectrum of the fluctuations of stress ratio at residual state.

(DTFT) and Discrete Fourier Transform (DFT), according to the properties (discrete or continuous, periodic or aperiodic) of signals they are dealing with. With the development of computer technology, the DFT and the algorithm for its fast calculation (FFT) are now easily performed with a PC while the others still remain as theoretical concepts (Smith, 1997). The output of the DFT is a set of numbers that represent amplitudes assigned to a set of sine and cosine waves with a series of frequencies, often shown on a frequency spectrum. By doing so, the DFT converts signals in the time domain into signals in the frequency domain. Though this technique is widely used in digital signal processing (DSP) in the field of mechanical engineering, electrical engineering and computer sciences (Smith, 1997), it is rarely used in geotechnical engineering or related research.

As the signals (stress ratio and vertical displacement) in this study is in the length (shear displacement) domain rather than time domain, the DFT can just convert them into signals in the domain of length-related frequency rather than real frequency with the dimension of reciprocal of time (unit of 1/sec, Hz). We define length frequency (*LF*), which has the dimension of reciprocal of length (unit of 1/mm, lHz), to express the number of times the wave of stress ratio and/or vertical displacement occurs within a unit shear displacement. Based on this definition, the wave length becomes of the reciprocal of length frequency (1/*LF*), without direct relation with the wave velocity. Taking fluctuations of stress ratio and vertical displacement (as displayed in Fig. 4.16) as input, the DFT produces a length frequency spectrum, presenting amplitudes assigned to a set of sine/cosine waves with a series of length frequencies (as displayed in Fig. 4.18).

As the DFT requires $2^n$ data points at equal interval, 4096 pairs of data, i.e. stress ratio and horizontal displacement, were cut out over the residual state of test 2III (Sample 2, 200 kPa, 0.06 mm/sec). After the selected data were decomposed by the DFT, the length frequency spectrum was produced as shown in Figure 4.18, illustrating details of

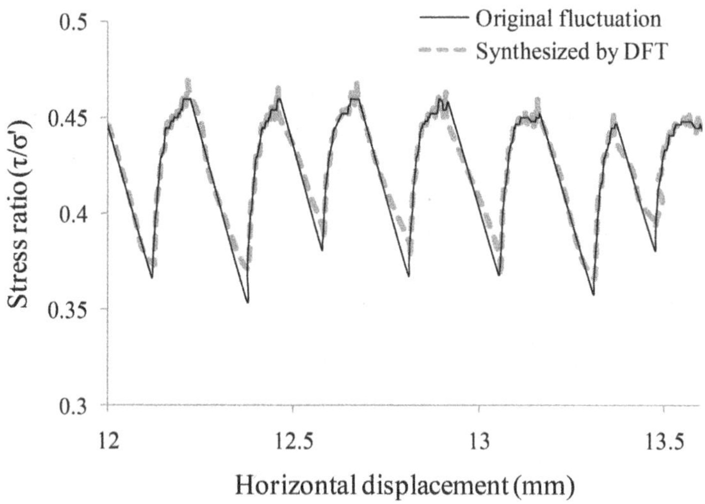


*Figure 4.19* Fluctuation of stress ratio can be reproduced by the inverse DFT procedure.

the frequency content of stress ratio along shear displacement. We can easily observe that there are three groups of outstanding spectral peaks in the ranges of 3–6 lHz, 12–18 lHz and 28–38 lHz, respectively. When performing the inverse DFT calculation (sometimes, called synthesis) by taking only the pronounced length frequency bands as input, the result from the backward calculation fits the original signals with negligible differences (Fig. 4.19). This indicates that the fluctuation along the shear displacement could be described by the combination of simple sine/cosine waves with the length frequencies in these bands as shown in Figure 4.20. The wave length, which is equal to the reciprocal of length frequency as defined above, of the fluctuation is approximately 0.25 mm when the most pronounced spectral peak (4 lHz) in Figure 4.18 is considered. This value is in good agreement with the wave length ($0.238 \pm 0.033$ mm) which was manually read out from curves of stress ratio and vertical displacement against horizontal displacement. This also reinforces the fact that the DFT is capable of decomposing the fluctuation and the most outstanding spectral peak can represent main characteristics of the fluctuation.

The success in decomposing fluctuations of measured shear stress also illuminates the way for the formulation of the energy system during the shearing process. Considering the sheared specimen during residual state as two rigid blocks, their relative movement could be divided into two, the horizontal advance and the vertical fluctuation. Therefore, the amount of energy used during this process can be thought of being composed of kinetic energy ($E_k$) pushing the upper block of the specimen laterally against friction and the wave energy ($E_w$) maintaining the fluctuations. The producer of the energy is only the motor driver pushing the upper shear box and the work it does can be expressed as the product ($W$) of measured horizontal force ($F(t)$) and the horizontal displacement ($\gamma(t)$), both of which are the function of time during the shearing process. Thus, the equations for this work-energy system can be written as follows:

$$W = E_k + E_w \tag{4.1}$$

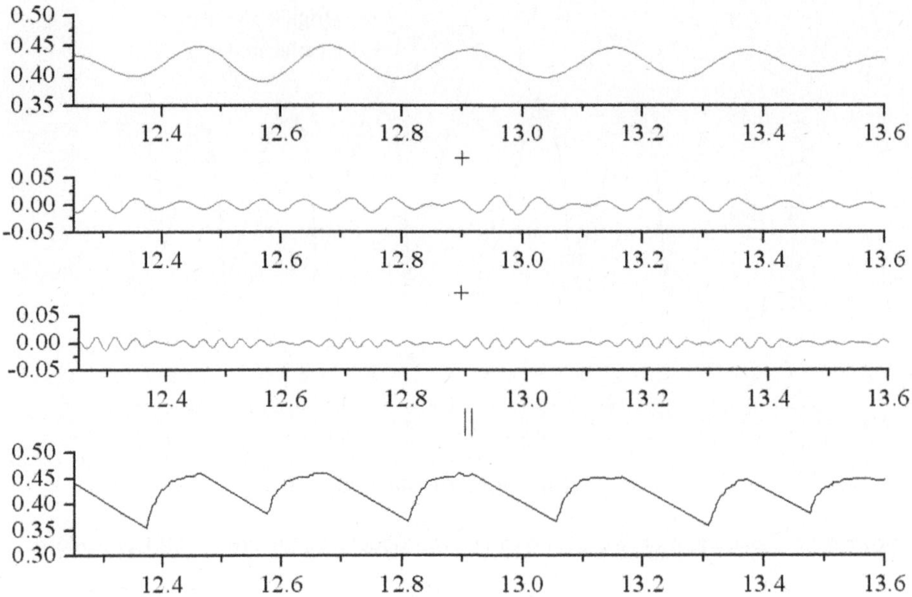


*Figure 4.20* The scheme of the DFT to decompose the fluctuation of stress ratio.

$$W = F(t) \times \gamma(t) \tag{4.2}$$

$E_k$ is equal to the product of friction force ($f_r(t)$) and the horizontal displacement ($\gamma(t)$) as

$$E_k = f_r(t) \times \gamma(t) \tag{4.3}$$

According to the standard wave equation

$$y = A\cos(wt + \phi) \tag{4.4}$$

where $y$ is the vertical displacement from the neutral position, $A$ is the wave amplitude, $w$ ($=2\pi f$) is the angular frequency, $f$ is the length frequency and $\varphi$ is the initial phase angle. At the moment $t$, the energy ($E_w$) of the particle subjected to the wave is

$$E_w = mA^2w^2\sin^2(wt) \tag{4.5}$$

Thus, Equation 4.1 can be written as

$$F(t) \times \gamma(t) - f_r(t) \times \gamma(t) = mA^2w^2\sin^2(wt) \tag{4.6}$$

where $m$ is the mass of the particle. To get the average energy of the wave over an entire wave length, we take the integral of both sides over the time range from 0 to $T$ (the wave period).

$$\frac{1}{T}\int_0^T (F(t)r(t) - f_r(t)r(t))dt = \frac{1}{T}\int_0^T mA^2\omega^2 \sin^2(wt)\, dt \tag{4.7}$$

where $r(t)$ is the shearing rate. Using the summation formula to substitute the left hand side of the above equation gives

$$\underset{\Delta t \to 0}{\text{limit}} \left( \frac{1}{T}\sum_{i=1}^{n} F(i)r(i)\Delta t - \frac{1}{T}\sum_{i=1}^{n} f(i)r(i)\Delta t \right) = {}^1\!/\!_2 mA^2\omega^2 \tag{4.8}$$

where $n$ is the number of the data points within a period $T$, the time interval $\Delta t = T/n$. For simplicity, the three parts in Equation 4.8 are designated by the notations $\overline{F_T}$, $\overline{f_T}$ and $\overline{V_T}$.

$$\overline{F_T} - \overline{f_T} = \overline{V_T} \tag{4.9}$$

Here, we define $\overline{F_T}$, $\overline{f_T}$ and $\overline{V_T}$ as the T-average work of horizontal force, the T-average work of friction and the T-average energy of wave, respectively.

Inserting the numerical data, taken from a wave-length range over the residual state of test 2III, into Equation 4.8, gives $\overline{F_T} = 0.0585$ and $\overline{V_T} = 0.0061$, and thus, $\overline{f_T} = 0.0524$. In this case, the work by horizontal force is 0.0585, and 89.6% of this value is dissipated by friction (0.0524), and 10.4% is dissipated by fluctuation (0.0061). This indicates that even during the residual state of specimens with coarse particles, the volume change still play an important role in contributing to the total shear strength. Taylor (1948) divided peak shear stress ratio $\tau/\sigma'$ into two components, interlocking ($\delta y/\delta x$) and friction $\mu$, that is $\tau/\sigma' = (\delta y/\delta x) + \mu$. With this model, he found that at the peak point, the total shear stress of 1.94 tons per sq ft consisted of the frictional strength (74%) and the strength arising from resistance to volume increase (dilation) (26%).

### *Proposed shearing model for granular material*

On all the plots in Figures 4.14 to 4.16, there appears to be the same tendency for the fluctuation development along the displacement. In the period between point A (the peak) and point B (maybe the critical), the amplitude continuously grows up with shearing. This may be due to the rearrangement of particles and exchange of particles between the shear zone and its vicinity during shearing. Shimizu (1997) indicated that before the peak stress occurs the strains tend to localize in the end zones of the potential shear zone, and after the peak, strain increments develop almost uniformly within the shear zone. The development of shear zone is a progressive process: after the peak (point A), the potential shear zone becomes more active with relative movement of particles, resulting in a looser layer. During dilatancy, large particles in the vicinity of shear zone easily penetrates into the relatively loose layer since the energy needed to penetrate into the moving materials is the least and small particles inside the shear zone tend to drop into the voids built with coarse skeleton (Fukuoka *et al.*, 2006; Igwe *et al.*, 2007; Fukuoka *et al.*, 2007). In this way, coarse particles tend to concentrate in the shear zone while small particles get enriched outside the shear zone. This phenomenon was called segregation by recent researchers (Wafid *et al.*, 2004; Fukuoka *et al.*, 2007) who performed ring shear tests on sand and glass beads. The particle size distribution analyses on slices cut parallel to the shear zone showed that the slice containing the shear zone has 30% higher coarse fraction than other slices after shearing a sand specimen at 22 kPa normal stress (Fukuoka *et al.*, 2007). The so-called segregation ceases at point B where a steady shear zone forms as a relatively regular but uneven layer with neighboring resistant coarse particles entrapped in or drawn into the matrix of this layer, and thus not causing significant dilatancy. Hereafter, the inner structure of the specimen remains steady and the upper block moves forward with constant wave amplitude no matter how long it is sheared.

The characteristics (amplitude and wave length) of the fluctuations at increasing normal stress levels (Figs. 4.14 to 4.16) have important differences, though both the wave length and amplitude increase with the normal stress (Fig. 4.17(a–c)). At the normal stress of 50 kPa, many small-scale fluctuations are superimposed on the main ones. These small-scale fluctuations gradually disappear at higher normal stresses. This can be attributed to penetration of a greater number of coarse particles at higher normal stresses during the process from point A to point B. To accommodate the penetration of coarse particles, small particles are more susceptible to be pushed out of the shear zone under higher normal stress, leading to high concentration of coarse particles and therefore a rougher zone and stronger fluctuations. Figure 4.21(a–c) gives a magnified view of the conceptual model of the shear zones produced under different normal stresses. Higher normal stress causes greater amplitude and wave length of the fluctuation, indicating a shear zone with coarse particles scattering in the shear zone with farther spacing. On the contrary, lower normal stress leads to lower amplitude and wave length, indicating a relative thin shear zone with coarse particles scattering in the shear zone with close spacing.

On the basis of the analysis presented above, the shearing process in granular materials can be said to take place in four stages. The first stage exhibits contraction which may be due to the change of stress field caused by applied shear force. This force modifies the skeleton of the solid particles, which was formed during the previous consolidation period. With this distortion of stress field, the particles, especially at the boundaries of the shear zone, find their ways to move into the nearby voids. This stage ceases when granular interlocking prevents further contraction (Taylor, 1948).

In the second stage, i.e. dilation before the peak, the main movement of particles occurs to overcome the interlocking leading to volume increase; the peak occurs when the particle interlocking reaches its climax. After this, the shearing enters into the third stage where the slip surface starts to develop with small-size particles being pushed out of shear zone to accommodate the larger particles penetrating the shear zone. In this stage, normal stress level has a profound influence on the formation of shear zone by regulating the concentration of coarse particles, and thus producing different fluctuation patterns as shown in Figure 4.21(a–c). At the end of the third stage, rigid blocks are formed. From the macroscopic point of view, the main style of movement becomes sliding along a wavy surface, though the particle interlocking still plays an important role in the magnitude and pattern of shear stress.

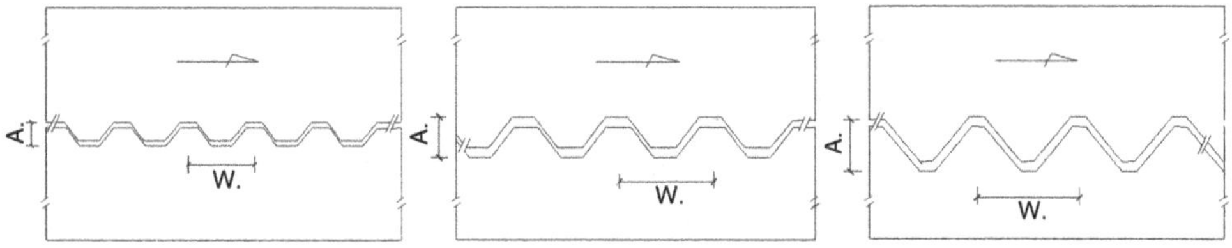


*Figure 4.21* The conceptual models of shear surface under different normal stress levels: (a) 50 kPa; (b) 100 kPa; and (c) 200 kPa.

Movement of particles into voids leading to local contraction, and their interaction leading to local dilation when they come into contact, coexists and moderate each other throughout the shearing process. Therefore, the observed stress fluctuations are the net effect of these two processes across the shear zone. During the first stage of shearing, the former is dominant. At the lowest point of volumetric strain, these two processes neutralize; after that dilation overtakes the former until the peak strength is reached; subsequently, contraction reappears and operates together with dilation. When the slip zone is ultimately formed these two processes reach a dynamic equilibrium.

## 4.3.4 Composite soils

As explained above, the samples composed of artificial granular materials (glass sand, glass beads) exhibit the similar shear behavior as dense sand, regardless of the normal stresses imposed on them and the shearing rates (Fig. 4.6(a–e)). Similarly, Samples 1 (kaolin) and 26 (kaolin and silt) exhibit the standard shear behavior of normally consolidated clay, gradually mobilized to residual state with slight brittleness (Fig. 4.6(a–e)). In contrast to these, the mixtures of fine and coarse fractions present a complicated system of shear pattern, which is seemingly influenced, to different extents, by factors such as Atterberg limits, gradation, normal stress, particle shape and shearing rate. In the classical soil mechanics and the critical state soil mechanics, there are two basic types of shearing patterns, which are often encountered in the laboratory tests. The first type is observed when dense sand or overconsolidated clay is sheared, showing contraction, dilation and then leveling off. The second type occurs when loose sand or normally consolidated clay is sheared, showing contraction and then leveling off. For simplicity, hereinafter, we call the former pattern as dilation and the latter as contraction.

### *Tests of KG mixtures*

As shown in Figure 4.6(a–e) the reduction of shear strength (brittleness) occurred after peak point for Samples 1, 7 and 8. All the tests on these samples exhibit contraction during shear, regardless of the magnitude of normal stress. In these clay-rich samples, the reduction of strength from peak to residual and the contraction pattern is considered to have resulted primarily from the formation of extremely thin layers of fine particles orientated in the direction of shear; these particles would originally have been in a random state of orientation and must therefore have had a greater resistance to shear than when they became parallel to each other in the shear direction (Skempton, 1964;

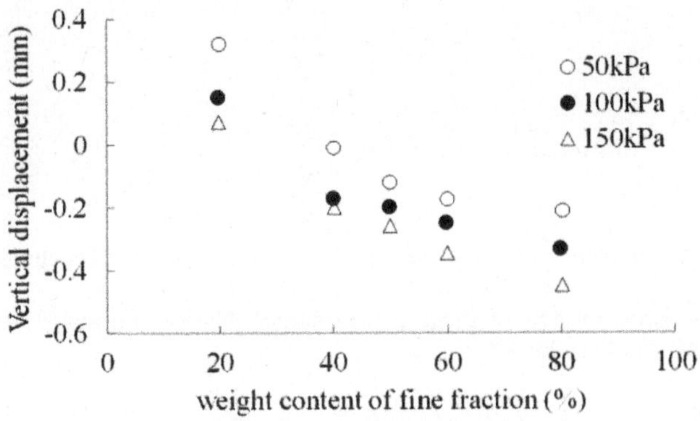


*Figure 4.22* Correlation between vertical displacement and fine fraction for the tests on KG mixtures.

Kenney, 1977; Lupini *et al.*, 1981). Though interlocking of coarse particles could not have been completely avoided, the high proportion of fine fraction (more than 60%) allowed the sliding of kaolin to prevail, and particle interlocking to be weak.

The shearing on Sample 11 also shows strong brittleness, which is consistent with the observed dilation. In this case, the low fine fraction (20%) makes the residual strength essentially that of the rotund particles alone, though kaolin particles, perhaps those at the contacts between the rotund particles, results in a reduction in strength. In the case of dense sand, as Taylor (1948) demonstrated, the brittleness is induced mostly by particle interlocking, as it consumes about 1/4 work of measured shear stress, approximately equal to the difference in strength between peak and residual, to make the specimen dilate, producing a distinct peak point. The high proportion of coarse fraction (80%) makes particle interlocking prevail and diminishes the effect of sliding of kaolin.

Specimen 9I shows non-brittle dilation during shearing, while Specimens 9II and 9III show non-brittle contraction. Testing results from Specimens 10I and 10II exhibit non-brittle dilation while those from Specimen 10III show non-brittle contraction. The intermediate proportion of fine fraction (50% in Sample 9 and 40% in Sample 10) makes it possible for shear surface to contain slickensides, though locally disrupted by coarse particles. This means the sliding and particle interlocking coexist. The saved energy due to formation of discontinuous slickensides compensates for the energy needed to overcome particle interlocking, thus disabling appearance of a distinct peak and leading to a non-brittle behavior.

Plots of vertical displacement against proportion of fine fraction (Figs. 4.22 and 4.23) indicate that the vertical displacement may be a function of both gradation and magnitude of normal stress. Also referring to Figure 4.6(a–e), with the increase of fine fraction, the vertical displacement pattern gradually transforms from dilation to contraction; if with the same fine fraction, e.g. 50% for Sample 9, the tendency of shear pattern transformation was held back by the increase of normal stress and it is found that the lower is the normal stress, the higher is the potential to exhibit dilation. For example, test 9I shows dilation whereas 9II and 9III remain of contraction pattern,

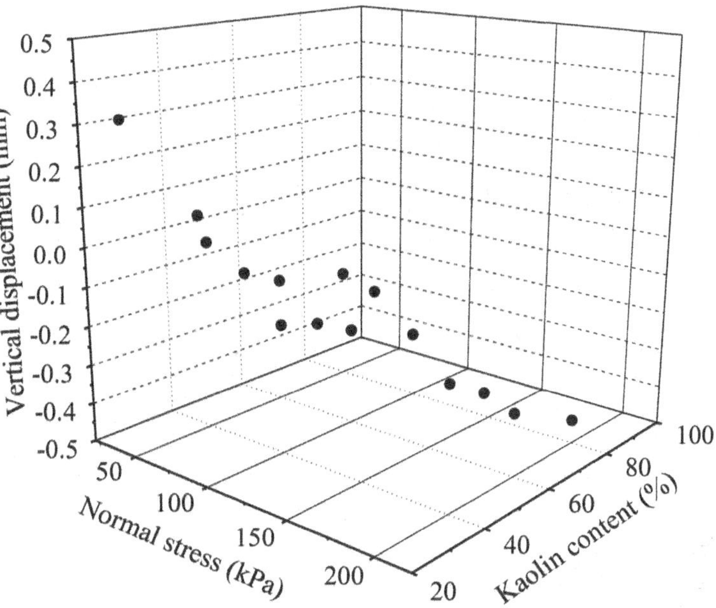


*Figure 4.23* Dilation pattern/vertical displacement is a function of both normal stress and the fine fraction for the tests on KG mixtures.

though the gradation of these three specimens is the same. Accompanying the vertical displacement, it seems that the brittleness follows a curved trace depending on the proportion of fine fraction, reaching a minimum (non-brittle) when this proportion is about 40% to 50% and a climax when it is either 100% or 0%.

After removal out of the shear box, sheared specimens derived from Samples 1 (100% kaolin), 7 (80% kaolin) and 8 (60% kaolin) were easily split into two parts, regardless of the magnitude of normal stress. When the fine fraction decreased to 50% in Sample 9, the specimen 9III, sheared at 200 kPa, could be split along the shear zone with some effort, whereas Specimens 9I and 9II, sheared at 50 kPa and 100 kPa respectively, were very difficult to split. In the cases of low fine fraction either of 40% in Sample 10 or of 20% in Sample 11, it was extremely difficult to split when the specimens were wet, regardless of the magnitude of normal stress. In order to examine the features of shear zone, these specimens were carefully removed from the shear box and air dried at 20°C, and the specimens were split with much care. The photos of the shear zone were shown in Figure 4.24(a–f). The graduated scale on the photos is readable to 1.0 mm.

The slickenside gradually disappears with the decrease of fine fraction and the shear surface gets rougher. Shear surface of Specimen 1III (100% kaolin) is reflective, even and well slickensided in the direction of shear; the shear surface of Specimen 7III (80% kaolin) is characterized with fluctuating slickensides caused by the penetration of coarse particles. In the case of Specimen 8III (60% kaolin), slickenside could not spread over the entire shear surface and the local congregation of coarse particles

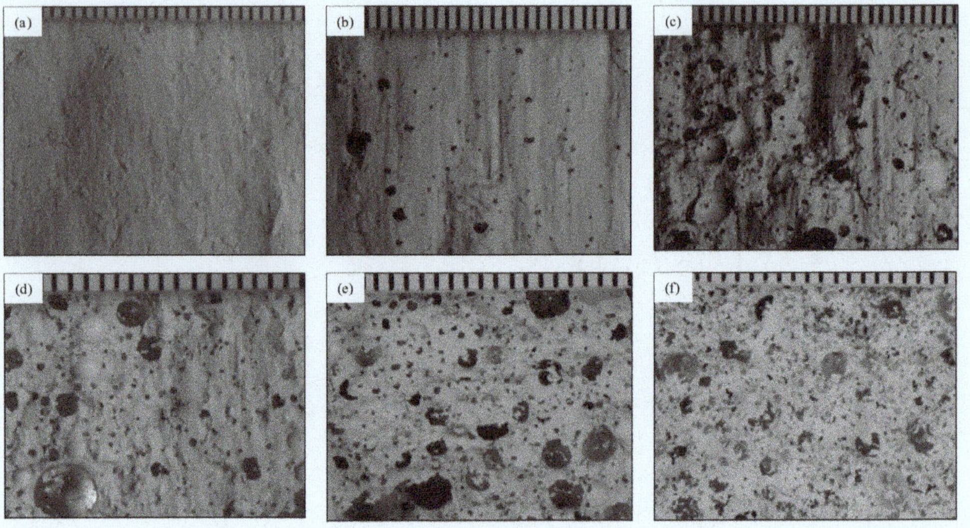

*Figure 4.24* Shear zones of: (a) Specimen 1III; (b) Specimen 7III; (c) Specimen 8III; (d) Specimen 9III; (e) Specimen 10III; and (f) Specimen 11III.

with surrounding fines was observed, which caused deep grooves (or coarse striations) in the shear zone. The shear surface of Specimen 9III (50% kaolin) presents poorly developed and discontinuous slickensides. There was no evidence that the Specimens 10III (40% kaolin) and 11III (20% kaolin) contain a discontinuously striated shear surface. The shear zone was also found to increase in thickness with decreasing fine fraction. The series of photos in Figure 4.24(a–f) illustrates that 50% fine fraction may be the threshold for this type of mixtures to indicate whether or not a distinct shear zone could develop during shearing. This is in good agreement with what has been found by Lupini *et al.* (1981), when he tested the mixtures with various proportions of mica and sand and that of London clay and Happisburgh sand.

What is found here by observation of the shear zone structure coincides well with the pattern of vertical displacement and brittleness property discussed in previous section. Sliding of fine fraction and interlocking of coarse particles coexist during shearing of the mixture. In the case of high proportion of fine fraction, sliding prevails, causing contraction. On the contrary, the particle interlocking dominates if the proportion of fine fraction is low, causing dilation. At lower normal stress, it is easy for round glass particles to overcome the resultant force, induced by normal and shear stresses, and ride over each other when they come into contact, causing dilation. In this case, slickensided shear surface hardly forms and the shear surface either has poorly developed and discontinuous slickensides (Fig. 4.25b) or just a thick pile of materials (Fig. 4.25a). In contrast, at higher normal stress, when coarse particles come to contact, the roundness enable them to be pushed into the matrix, as the matrix is not stiff enough to confront the higher resultant force, introducing contraction. In this case, a thin shear zone with fine particles forms with the shearing process (Fig. 4.25c).

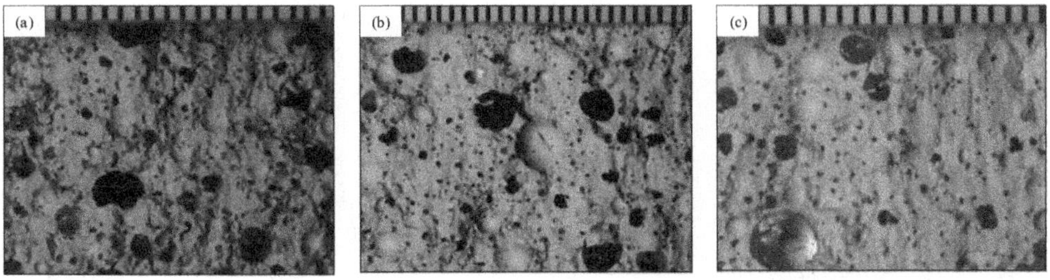


*Figure 4.25* Shear zones of: (a) Specimen 9I; (b) Specimen 9II; and (c) Specimen 9III.

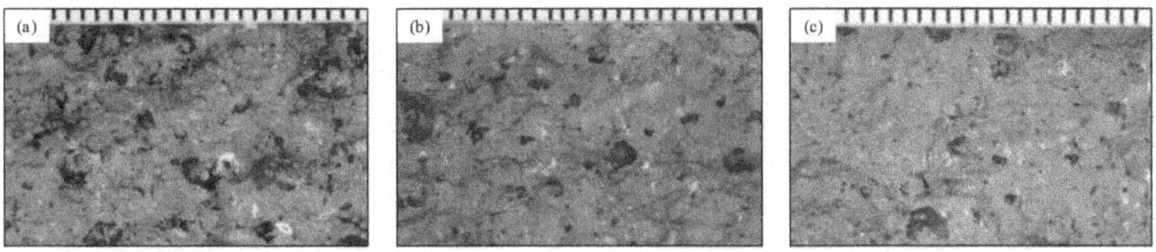


*Figure 4.26* Shear zones of specimens of FG mixture: (a) Specimen 13I; (b) Specimen 13II; and (c) Specimen 13III.

### *Tests of FG mixtures*

In general, for tests on samples of FG mixtures, the tendency for the transformation of dilation pattern is similar to what is observed from the tests of KG mixtures. As mentioned above, this transformation is controlled by both the magnitude of normal stress and the gradation. However, the samples of FG mixture are more prone to exhibit dilation than samples of KG mixture, probably due to the contribution of silt which does not exist in the samples of KG mixture. For example, Specimen 8I (40% coarse and 60% kaolin, 50 kPa) shows contraction while Specimen 13I (40% coarse and 60% mixture of silt and kaolin, 50 kPa) shows dilation, though the coarse fraction remain the same. Figure 4.26(a–c) shows the structure of shear surfaces of Specimens 13I, 13II and 13III. It is clear that the shear surface is getting flatter with increasing normal stress. The sliding movement prevails at high normal stress while the turbulent movement is the main mode at low normal stress. According to Figure 4.6(a–e), the proportion of fine fraction (kaolin and silt) in Sample 13, 60%, looks as the threshold level, at which the sliding neutralizes the particle interlocking, showing non-brittle behavior, whereas the other samples, with either less or more fine fraction, show brittleness.

### *Tests of FN mixtures*

At 200 kPa, samples of FN mixture become more prone to dilate with increase of coarse fraction. The angularity and roughness of coarse particles (river sand and crushed granite gravel) reduce the ease to push them into the fine matrix and cause interlocking of particles, resulting in large dilation. The interlocking may also cause particle segregation, making coarse particle accumulate in the shear zone. Splitting the sheared

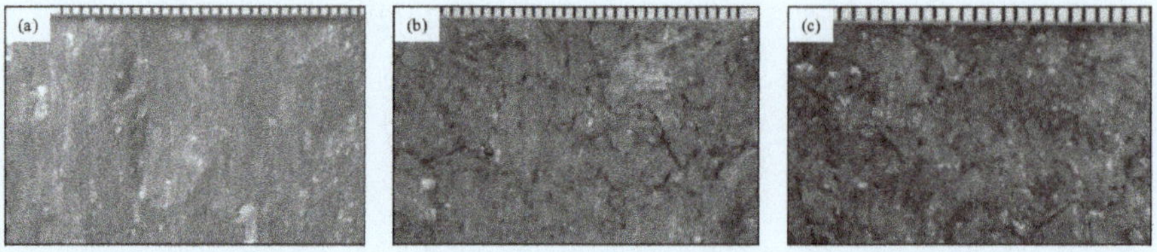

*Figure 4.27* Shear zones of specimens of FN mixture: (a) Specimen 26VI; (b) Specimen 19VI; and (c) Specimen 21VI.

specimens in this category, sheared at 200 kPa, was unsuccessful, both when wet immediately after removal from the shear box and when air dried in the lab. The photos in Figure 4.27(a–c) was taken from the shear zones of Specimens 27VI (100% fine), 19VI (80% fine) and 21VI (60% fine), which are sheared at 400 kPa. The specimens with less than 60% fine fraction could not be split along the shear surface. Visual examination indicates that the thickness of shear zone of the specimen with 100% fine fraction is approximately 0.5 mm. The thickness increases up to approximately 2.0 mm with decreasing fine fraction down to 20%. The photos in Figure 4.27(a–c) show that the shear surface gets rougher and more porous with the decrease of fine fraction. Specimen 26VI has a flatter surface with continuous striations; Specimen 19VI shows a relatively flatter surface with some discontinuous slickensides disrupted by coarse particles, whereas Specimen 21VI presents weak evidence of short slickensides scattered on the surface.

### 4.3.5 Residual shear modes

In the frequently cited paper, Lupini *et al.* (1981) put forward three modes of residual shear behavior; a turbulent mode, a transitional mode and a sliding mode. He indicated that the mode depends upon dominant particle shape and on the coefficient of interparticle friction. These two factors actually overlap as the coefficient of interparticle friction has strong positive correlation with the particle shape. Since the tests were only on the mixtures of sand and clay, the limitation of the range of particle size made this conclusion applicable only for soils with particles less than 2.0 mm (Lupini *et al.*, 1981). According to the testing results from present study, these three modes are to be modified and extended for application to a wide range of soils in terms of the maximum particle size involved. The visual examination on the structure of shear zones in previous sections suggests that the residual shear mode of specimen behaviors are modified by the cooperation of many factors, such as the gradation (PSD), magnitude of normal stress ($\sigma'$) and particle shape of coarse particles. These factors determine which one of the two basic modes of particle response will dominate during shear, the orientation (sliding) of fine particle or the interlocking of coarse particle. The dominant overall particle response will in turn determine the residual shear mode.

Kenney (1977) recognized the importance of the proportions of clay mineral particles to massive particles (coarser than clay size) present in a mixture. He defined the relative residual strength as

$$R_\varphi = \frac{\tan\varphi'_R - (\tan\varphi'_R)_C}{(\tan\varphi'_R)_M - (\tan\varphi'_R)_C} \tag{4.10}$$

where $\phi'_R$, $(\phi'_R)_C$ and $(\phi'_R)_M$ are the residual friction angles of the mixture, the clay minerals and the massive particles (coarser than clay size) respectively. He also defined a volume ratio $r_{vc}$ as

$$r_{vc} = \frac{\text{volume of clay minerals and water}}{\text{total volume of soil}} \tag{4.11}$$

and produced good correlations between $R_\varphi$ and $r_{vc}$ when he tested sand mixed with various clays. Lupini *et al.* (1981) argued that there is an ambiguity in this definition, in which Kenney considered the volume of water to be that associated with the clay minerals. This consideration makes $r_{vc}$ to be zero in fully saturated sand. Instead, Lupini *et al.* (1981) defined the granular void ratio $e_g$ as

$$e_g = \frac{\text{volume of platy particles and water}}{\text{volume of rotund particles}} = \frac{r_{vc}}{1 - r_{vc}} \tag{4.12}$$

The parameter was widely used by previous researchers when they performed tests on soil mixtures (Georgiannou *et al.*, 1990; Vallejo, 2001). The ratio $e_g$ can be calculated in an easy way, $e_g = (V_0 - V_g)/V_g$, where $V_g = M_g/R_g + M_s/R_s + M_m/R_m$, $M_g$, $M_s$, $M_m$, $R_g$, $R_s$ and $R_m$ are the mass and specific gravity of gravel, sand and silt, respectively, provided there is no particle breakage during shearing (Vallejo, 2001). Since the area of the sample in the shear box is known, one only needs to measure the height of the sample after consolidation ceased to obtain the total volume of the sample, $V_0$.

The relative residual strength $R_\varphi$ was plotted against the granular void ratio $e_g$ in Figure 4.28. The data shown in this figure is only for the tests conducted at normal stress of 200 kPa in order to avoid the interference of other factors. The results for the mixtures tested by Kenney (1977) and Lupini *et al.* (1981) were also plotted. For each mixture, Lupini *et al.* (1981) marked a range of $e_g$, over which the soil displayed transition in shear behavior from sliding to turbulent mode. He suggested that the behavior the sand-London clay mixture exhibits could be used as a guide to the type of residual behavior of natural soils due to its common gradation. Over the range of $e_g$ from 1.0 to 2.2, the soil (<2.0 mm) behaves in transitional mode; if $e_g$ is less than 1.0, turbulent shear would occur and if $e_g$ is greater than 2.2, sliding shear would occur. The $R_\varphi - e_g$ trend line obtained here shows good coincidences with the shear zone structure presented in Figures 4.24 to 4.27 and the mobilized residual shear strength: samples with $e_g > 4.0$ have a well slickensided shear surface and a relatively low residual shear strength; samples with $e_g < 0.8$ usually have a reworked shear zone rather than a distinct shear surface and a high residual shear strength and samples with $e_g$ ranging from 0.8 to 4.0 have a relatively thick shear zone with a roughly striated shear surface. Based on the flexuous property of the relationship between $R_\varphi$ and $e_g$ and the shear zone structure, the $e_g$ range from 0.8 to 4.0 (between two thick lines) could be distinguished, over which the shear behavior changes from sliding to turbulent.

Figure 4.29 plots $R_\varphi$ against $e_g$ for the KG mixture at three levels of normal stresses, i.e. 50 kPa, 100 kPa and 200 kPa. The filled area marks the $e_g$ range, over which the soil exhibits transitional residual behavior, over the $S$ range sliding mode occurs and

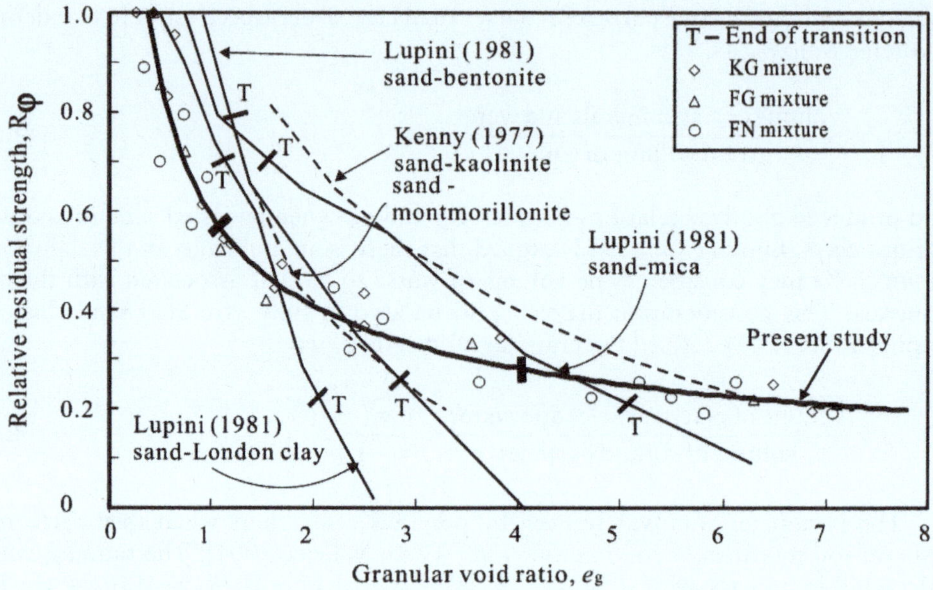


*Figure 4.28* Relative residual strength, $R_\varphi$, versus granular void ratio, $e_g$, for various soil mixtures.

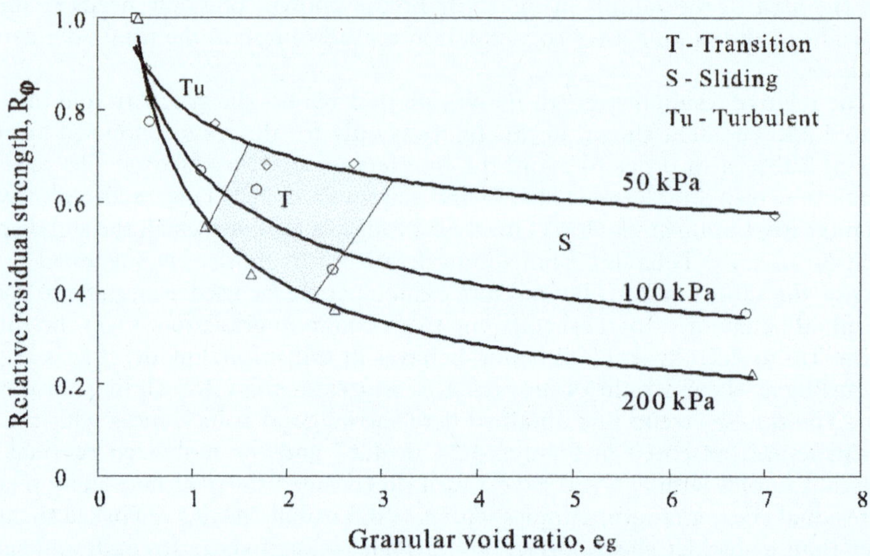


*Figure 4.29* Influence of normal stress on the transition of shear modes, shown in $R_\varphi - e_g$ space.

Tu range is for the turbulent mode. The inflexion section of each line show good agreement with what has been observed through the visual examination on shear zones. The normal stress imposes influence on the transition of shear modes. Sample 8, when subjected to shearing at 50 kPa (8I), stops the sliding mode and exhibits transitional

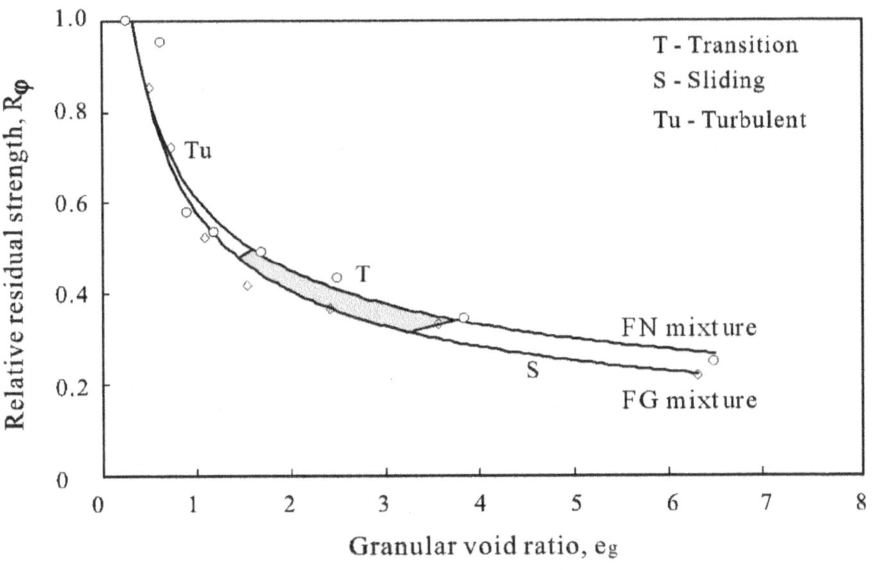


*Figure 4.30* Influence of particle angularity on the transition of shear modes, shown in $R_\varphi - e_g$ space.

mode with discontinuous slickensides and coarse particles within the shear zone. On the contrary, it shows sliding mode when sheared at 200 kPa (8III), with strongly slickensided shear surface. The same phenomenon was observed in Sample 9. Shearing at 200 kPa produced a relatively flatter shear surface, giving a residual friction angle of 26°, while 50 kPa produced a thick shear zone with a loose but turbulent structure, causing an increase in shear strength by 6°.

Figure 4.30 plots $R_\varphi$ against $e_g$ for the FG and FN mixtures at the same normal stress of 400 kPa. Again, the inflexion section of each line show good agreement with what has been observed through the visual examination of the shear zones. Samples 17 and 20 both have the same gradation with 70% fine fraction. However, when tested at 400 kPa, Sample 17 (FG) shows sliding behavior whereas Sample 20 (FN) shows turbulent behavior. This can be observed by examining the shear zone structure. The former has residual friction angle of 24° and the latter 28°. This is likely due to the coarse particles in FN mixture, being irregular and rough in shape, which provide high particle friction and strong particle interlocking as they cannot orientate themselves. The spherical and smooth glass particles in FG mixture, on the other hand, can be easily pushed into the fine matrix or easily slide over each other when in contact.

Based on the description and interpretation above, the characteristics of residual shear modes can be concluded. The soil exhibiting sliding shear mode contains a high proportion of fine and low-friction particles, causing a high value of $e_g$. It, in general, shows relatively low residual strength and the shear surface is easy to be exposed by splitting the sheared specimen. The high proportion of fine particles makes it possible to form a slickensided sliding surface with strongly orientated clay particles. The particle interlocking may lead to a deep striation within the shear zone. However, it is incomparable to the pronounced sliding due to fine particles. The samples with more than 70% fines in the three mixtures show this type of shear behavior. The soil behaving in sliding mode is likely to show slight brittleness with the formation of slickensided shear surface. The water content of shear zone is generally lower than that of other parts of the specimen due to the relative dense structure.

The turbulent mode tends to occur in the soil dominated by coarse particles, which cause a low value of $e_g$. This type of soil shows a relatively high residual strength and the shear zone is thick and difficult to be exposed by splitting. The dominating coarse particles introduce strong particle interlocking, which outweigh sliding occurring between the fine particles. The samples with more than 70% coarse particles in the three mixtures show this type of shear mode with dilation and strong brittleness. The water content of shear zone is generally higher than that of other parts due to the loose structure within the shear zone.

The transitional mode occurs in the soil which has moderate proportion of both fine and coarse fraction. The shearing produces a zone, rather than a single shear surface. This zone contains both discontinuous slickensides and coarse particles and is a product of interaction between sliding and particle interlocking. The behavior of the soil is sensitive to the magnitude of normal stress and particle shape. A high normal stress would cause it to behave by sliding while roughness and angularity would cause it to behave in turbulent mode. The soil generally shows contraction-dilation pattern and non-brittleness. The difference in water content between shear zone and other parts is almost negligible.

The magnitude of normal stress and particle shape both influence the transition of shear modes, to some extent. A high normal stress favors sliding mode while the roughness and angularity of coarse particles intensify the particle interlocking, and favors transitional and turbulent shear modes. Since the samples tested in this study covered a wide range of soil types in terms of gradation and particle shape, the $e_g$ boundaries (0.8, 4.0) distinguished from them would be a useful guide to the type of residual behavior of natural soils with corresponding fine fraction and particle size ranges.

## 4.4 INFLUENCING FACTORS

As indicated in the introduction section of this chapter, the tests in the present study were mainly designed to discover the influential factors and the extent to which each factor contributes. Continuing the discussion in the previous section, where much emphasis was given to the shear pattern, the shear zone structure and the residual shear mode, this section aims to demonstrate, quantitatively and individually, the relationship between the residual shear strength and the five influencing factors recognized: particle size distribution, Atterberg limits, particle shape, magnitude of normal stress and shearing rate.

### 4.4.1 Particle size distribution

The particle size distribution (PSD) has long been recognized as an important factor influencing the shear behavior of soils and has been extensively studied (Taylor,

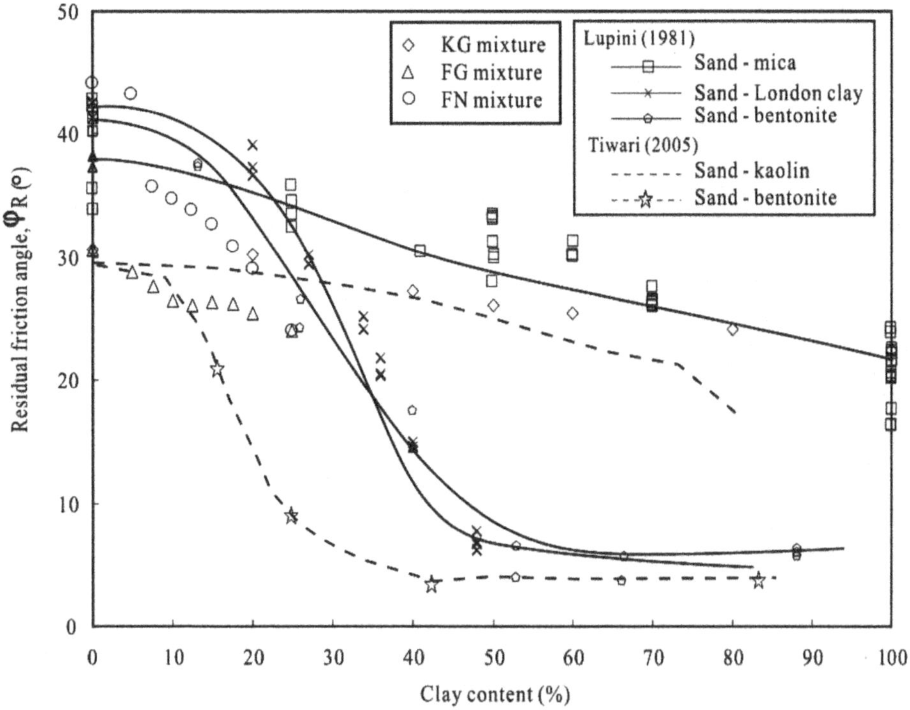


*Figure 4.31* Residual friction angle, $\varphi'_R$, versus clay content.

1948; Miller, 1957; Holtz, 1960; Skempton, 1964; Kenney, 1977; Lupini *et al.*, 1981; Kalteziotis, 1993; Irfan & Tang, 1992, 1995; Jovicic Coop, 1997; Ng & Chui, 2003; Dewoolkar & Huzjak, 2005; Simoni, 2006). The parameters which could possibly reflect the particle size distribution are distribution parameters $C_u$ and $C_c$, clay content, coarse particle content and various proportions of fractions, such as the ratio of clay content to coarse content and the ratio of gravel content to fine fraction, and so on. Among them, the clay content was most often used in literature (Lupini *et al.*, 1981; Irfan & Tang, 1992, 1995; Jovicic & Coop, 1997; Ng & Chui, 2003; Dewoolkar & Huzjak, 2005; Tiwari & Marui, 2005; Wen & Aydin, 2005).

The residual friction angle is plotted against clay content in Figure 4.31 together with test data from literature. Since the particle size distribution is taken into account by univariate analysis so as to identify and purify its influence on the residual strength, the relationship presented in this section is based on only the tests at normal stress of 200 kPa. By doing so, the complexity due to interference of normal stress may be avoided. The residual strength of samples of KG and FG mixture varies over the range from 24° to 31° while that of FG mixture varies over the range from 24° to 45°. The residual friction angle decreases with increasing clay content. The reduction in shear strength due to the increase in clay content should be attributed to the increased separation of the massive particles, if Skinner's (1969) conclusion, i.e. the residual

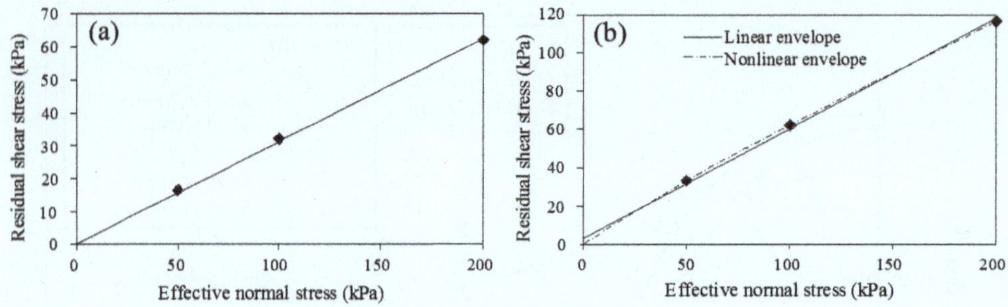


*Figure 4.32* Residual failure envelopes for samples: (a) Sample 1 (100% kaolin); and (b) Sample 11 (20% kaolin and 80% coarse fraction).

friction angle is not dependent on particle friction, is true. As illustrated by the visual examination of shear surface in Section 4.3.4, the mixture with a high proportion of clay fraction possesses a strongly slickensided shear surface and behaves dominantly in sliding mode. The mixture with a low proportion of clay fraction exhibits a thick shear zone, rather than a single shear surface. This zone rarely has slickensides and behaves in turbulent mode. When the mixture has moderate clay content, it possesses a shear zone that results from the neutralization of fine particle orientation and coarse particle interlocking, producing discontinuous slickensides and local increase of coarse particles.

The data presented by Lupini *et al.* (1981) for sand-mica mixtures and the data from Tiwari and Marui (2005) for sand-kaolin mixtures coincide with the results of this study. However, the data for sand-bentonite and sand-London clay mixtures are considerably different from the results of this study, except the clay content not exceeding 30%. Lupini *et al.* (1981) ascribed this difference to the particle size of clay minerals. The mica and kaolin are nearer in particle size to massive particles in the mixture than bentonite. This causes residual strength of clays to be similar to those of the mixtures. Substituting 75% clay by silt in the fine fraction introduces an increase of residual strength by 1°, indicating that the nature of coarse fraction (particle shape and/or size) would play an important role in the residual strength of the mixture. This is also recognized in this study as the FG and FN mixtures have a difference of approximately 9°. The latter shows more dependency on the clay/coarse fraction.

### 4.4.2 Normal stress

In the literature, the envelopes of effective residual shear strength often have zero (Bishop *et al.*, 1971) or small values (Lupini *et al.*, 1981; Skempton, 1985) of cohesion intercept. The residual shear strength results determined for the three types of mixtures in the present study does not always produce zero intercept residual strength envelopes. Instead, the linearly best fitted residual strength envelopes, in most cases, result in small values of intercept. As shown in Figure 4.32(a–b), Sample 1 has zero intercept envelop while Sample 11 (for example) has its intercept at approximately 1.2 kPa.

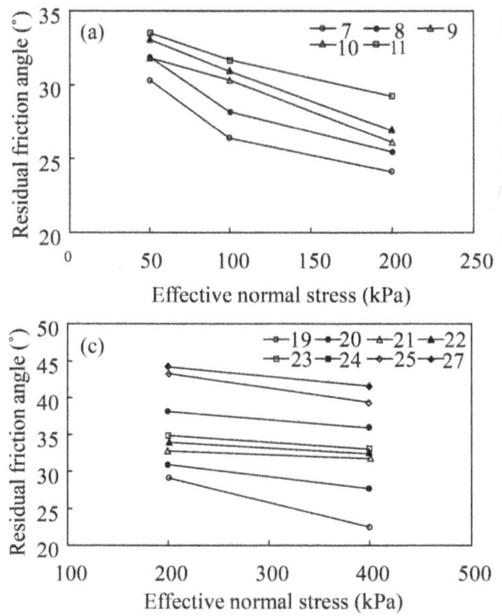


*Figure 4.33* Reduction in friction angle with increase in effective normal stress: (a) KG mixture; (b) FG mixture; and (c) FN mixture.

Other studies have concluded that the drained residual strength is normal stress dependent, i.e. the residual friction angle decreases with increasing effective normal stress (Kenney, 1967; Chandler 1969; Stark & Eid, 1994; Wan & Kwong, 2002; Tiwari *et al.*, 2005; Dewoolkar & Huzjak, 2005), with this effect being the most pronounced at low effective stresses. As previous researchers have performed, the residual friction angle was plotted versus the effective normal stress in Figure 4.33(a–c) for the three mixtures in present study. It can be clearly observed that all samples tested exhibit a decrease in the residual shear stress with increasing normal stress and this dependency is stronger over the lower range of normal stress, showing good agreement with literature.

Stark and Eid (1994) suggested that the stress dependent behavior of residual strength could be attributed to the rearrangement of platy clayey particles during shearing. They concluded that as the effective normal stress increases the edge-to-face interactions of platy clay particles tend to be converted to face-to-face interactions. Wan *et al.* (2002) argued Stark's explanation was not valid for soils without dominating clay fraction and the plastic junction theory proposed by Mitchell (1993) would be applicable to the soils with adequate sand and silt particles. The clay particles in the mixture coat the crystalline coarse particles, thus forming bridges which separate the coarse particles. This bridge is termed as the plastic junction and the frictional angle between bridged particles is smaller than without coating (Mitchell, 1993; Wan *et al.*, 2002). According to the plastic junction theory, an increase in the compression and shear stress causes the growth of plastic junction. The low friction caused by plastic junction reduces the interlocking of coarse particles and also enables coarse particles to be pushed into fine matrix, a relative smooth shear surface being formed. This could be observed by the visual examination of shear zones. Thus with the increase in normal stress, the residual shear stress decreases, showing a nonlinear rather than linear relationship when plotting in terms of Mohr-Coulomb envelope. Sample 1 (100% kaolin) shows non-dependency probably due to the fact that normal stresses used in this study do not involve the range less than 50 kPa, in which the cohesive soil is more susceptible (Kenney, 1967; Skempton, 1985; Tiwari *et al.*, 2005).

### 4.4.3 Atterberg limits

The residual shear strength has been long related to the plasticity index ($I_P$) in the literature and the $I_P$ value has proven to be an indicator of residual strength. Correlations developed by Voight (1973), Seycek (1978), Vaughan *et al.* (1978), Brandon *et al.* (2006), Mesri and Shahien (2003) and Tiwari and Marui (2005) are compared to the data from all three types of mixtures in Figure 4.34. In the cases of KG and FG mixtures, the residual friction angle decreases linearly with the increase of plasticity index and seemingly fits the boundary developed by Vaughan *et al.* (1978). At lower levels of $I_P$, the KG and FG mixtures show a large variation in $\varphi'_R$ for the same $I_P$. This is considered to be due to the influence of normal stress as discussed in previous section. The tests on FN mixture show good agreement with earlier research. Compared with KG and FG mixtures, the FN mixture exhibits stronger dependency on $I_P$. At 200 kPa, $\varphi'_R$ decreases from 44° to 25° with $I_P$ increasing from 0 to 28. This suggests the joint control of coarse particle shape and $I_P$, which mainly reflects the property of fine fraction of the mixture. The contribution of plasticity index is weak when the coarse particles are rounded and smooth with comparable interparticle friction to fine particles. In contrast, the irregular and rough particles would lose their noticeable interparticle friction once separated by fine fraction with high plasticity index. Thus the effect of $I_P$ is moderated by the particle shape. Also note the limited range of $I_P$ in this study, being less than 30%.

### 4.4.4 Particle shape

Sphericity, roundness, roughness and texture are some of the terms that were defined in many different ways and commonly used as the descriptors for the particle shape in the literature (Meloy, 1977; Barrett, 1980; Clark, 1987; Hyslip & Vallejo, 1997; Bowman *et al.*, 2001; Santamarina & Cho, 2004; Cho *et al.*, 2006). The recent research revealed that increasing angularity or decreasing roundness and sphericity leads to an increase in maximum $e_{max}$ and minimum $e_{min}$ void ratios and their difference $I_e$ (Shimobe & Moroto, 1995; Miura *et al.*, 1998; Dyskin *et al.*, 2001; Jia & Williams, 2001; Nakata *et al.*, 2001; Cubrinovski & Ishihara, 2002; Santamarina & Cho, 2004; Cho *et al.*, 2006). Cho *et al.* (2006) performed tests on natural and crushed sands to investigate the influence of particle shape on packing density, stiffness and strength and showed that increase in particle angularity also causes a decrease in stiffness and an increase in the critical state friction angle, proving the particle shape as an important soil property. It is also known that the presence of platy particles in granular soils leads to the decrease in packing density, stiffness and residual strength (de Graff-Johnson *et al.*, 1969; Lupini *et al.*, 1981; Hight *et al.*, 1998; Guimaraes, 2002; Cho *et al.*, 2006).

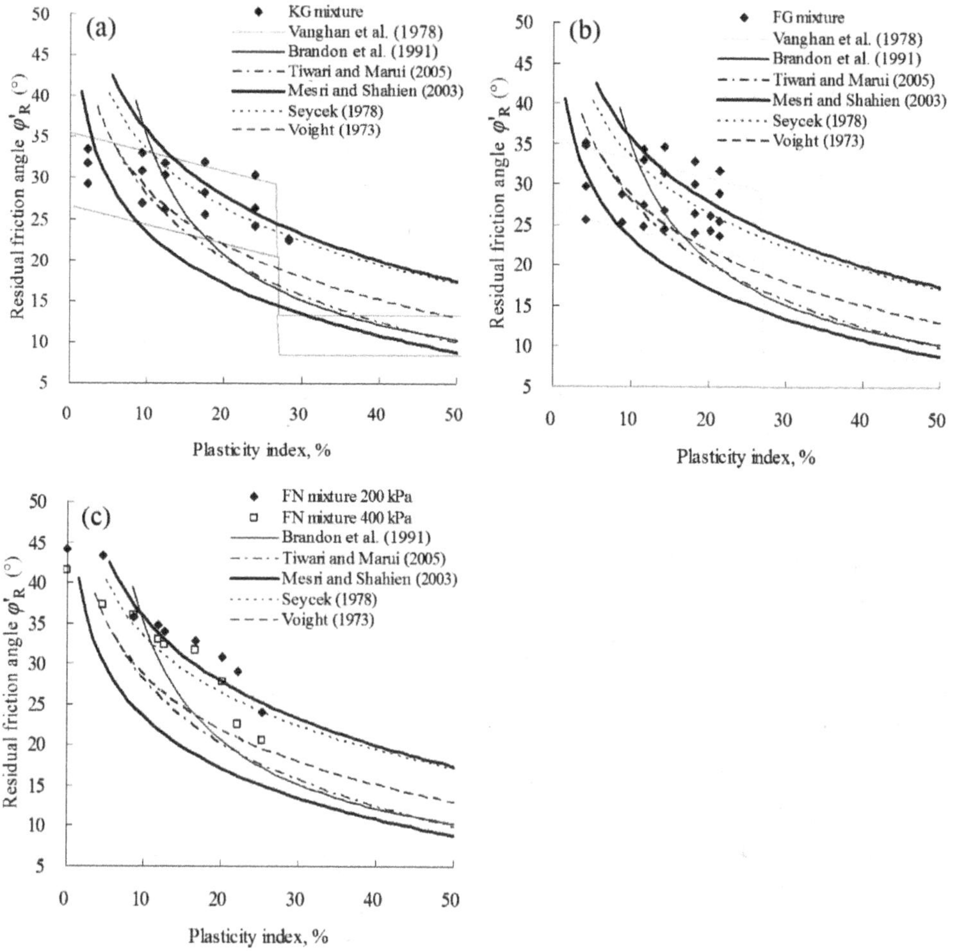


*Figure 4.34* Comparison of residual friction angle vs. plasticity index trends with correlations in the literature: (a) KG mixture; (b) FG mixture; and (c) FN mixture.

In the present study, a stereomicroscope (Olympus SZH10) equipped with an image capture and analysis system, analySIS, were employed to quantify the shape parameters of the soil particles. The stereomicroscope with its magnification capability of up to 140X enabled capturing the shape features of even fine silt fraction down to 0.002 mm. Referring to the 2D nature of the images (Fig. 2.6), the analySIS software uses the term "circularity" to quantify the "closeness" of a particle to a circle. Circularity is given by $4\pi A/P^2$, where $A$ is the particle area and $P$ is its perimeter. Thus a perfect circular/spherical particle has a circularity of 1.0 while a very "spiky" or irregular particle has a circularity value close to zero. The term "convexity" is used as a measure of surface roughness of a particle and is given by the ratio of the "convex hull perimeter" by the actual particle perimeter. Convexity also ranges from 1.0 for

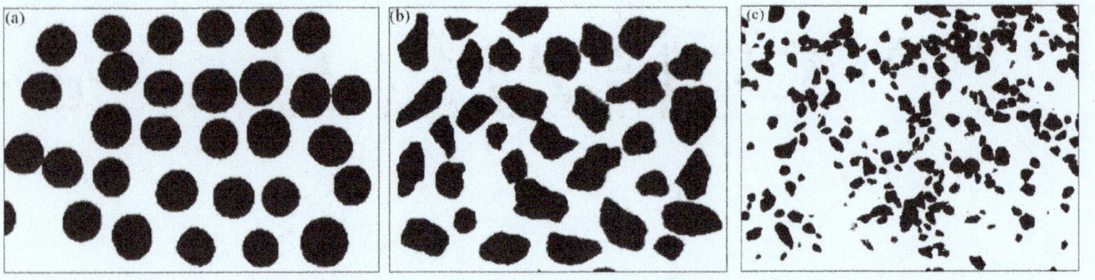


*Figure 4.35* Images of particles of: (a) rounded glass beads; (b) natural gravels; and (c) river sand.

a smooth shape to zero for a spiky or very irregular shape. The roundness of particles also can be expressed by elongation quantified by "1.0 – aspect ratio" or "1.0 – width/length". A particle symmetrical in all axes such as a circle or square will have an elongation value of zero whereas particles with small aspect ratios will have an elongation closer to 1.0.

The glass sand and glass beads in FG mixture are perfectly spherical and smooth, with both the circularity and the convexity being 1.0 and the elongation being zero (Fig. 4.35a). The natural coarse particles (river sand and crushed granite fragments) in FN mixture were measured, respectively, by means of the image analysis system (Fig. 4.35b and c). Clean and dried samples were sieved into sub-fractions as done in PSD analysis. The number of randomly selected particles from each fraction was large enough to determine the mean value for each parameter with variance less than 0.01. Shape parameters for each sample were obtained by weighted sum of the mean parameter values for each fraction (i.e. scaled by their corresponding weight percentages).

As described in the section on soil preparation, FG and FN mixtures have the same particle size distribution, but FG mixture uses rounded glass sand and glass beads as the coarse fraction while FN mixture river sand and crushed granite fragments. The difference in residual friction angle, $D_{\varphi}$, between FG and FN mixtures is plotted in Figure 4.36(a–c) against the corresponding differences in particle shape parameters, producing linear relationships. Increasing difference in elongation, $D_E$, leads to an increase in $D_{\varphi}$. However, increasing difference in circularity, $D_C$, causes a decrease in $D_{\varphi}$, so does the difference in convexity, $D_N$. Similar observations can be found in Chan and Page (1997) and Cho *et al.* (2006). Chan and Page (1997) also presented a relationship between critical state friction angle $\varphi_{cv}$ and particle roundness, $R$ as $\varphi_{cv} = 42 - 17R$.

Soil shearing involves particle rotation which prevails in loose soil, and contact slippage which prevails in dense soil. The residual strength is a result of evolution of anisotropy of soil fabric during shearing and reflects the ability of soil to develop internal forces and fabric anisotropy (Rothenburg & Bathurst, 1989; Thornton, 2000). It is reasonable to postulate particle angularity hampers particle rotation while roughness holds back slippage and both enhance dilatancy and anisotropy. In the test cases of this study, the presence of elongated particles in the shear zone makes it easy to work, if standing upward, as a stiff rod anchoring the upper and lower halves of sample together to prevent their relative movement. The linear relationships suggest the

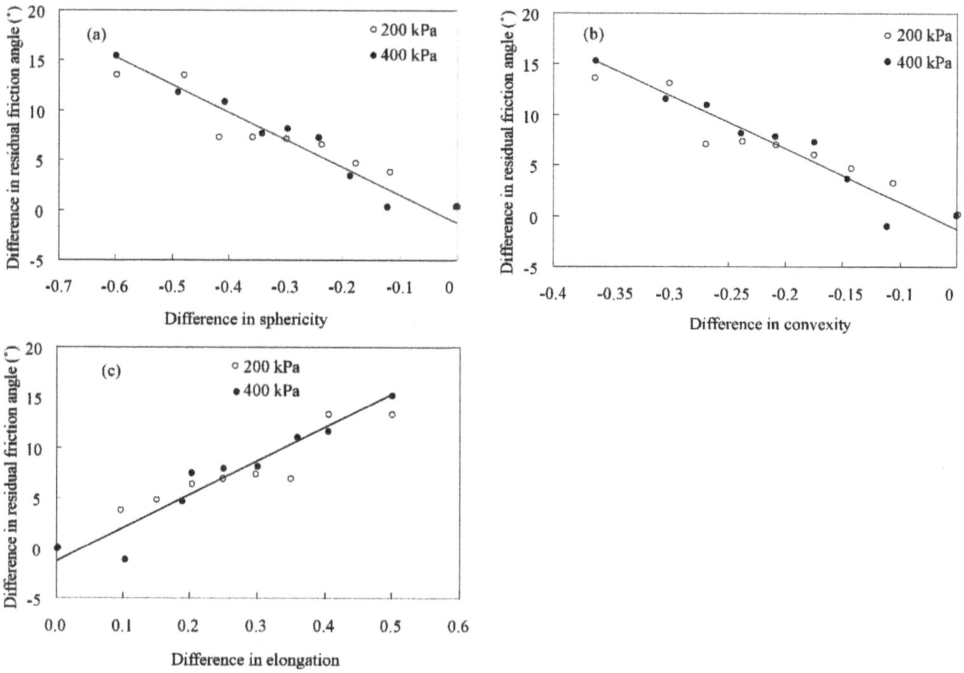


*Figure 4.36* Difference in residual friction angle is a function of the difference in particle shape parameters: (a) sphericity; (b) convexity; and (c) elongation.

participation of particle shape in affecting the residual friction angle. So far, though particle shape has been ignored in the process of soil classification and in the interpretation of residual shear strength, the tests in this study confirmed it to be a significant parameter that needs to be properly characterized and documented, in particular when soils contain sands or gravels.

### 4.4.5 Shearing rate

For investigation of the influence of shearing rate on the residual strength, the samples of KG mixture were sheared at three levels of shearing rates, 0.06, 0.6 and 6.0 mm/min. The samples of FG mixture were sheared at 0.06 and 0.6 mm/min. All tests for investigation of rate effect were conducted at the same normal stress, 200 kPa. The residual friction angle was plotted against the shearing rate in logarithmic scale in Figure 4.37(a–b). In the case of KG mixture, when the content of kaolin is greater than 40% increasing the shearing rate leads to a decrease in residual strength. The average reduction in the residual strength is about 1.7° for Samples 7 to 10. Residual strength of samples with kaolin content less than 40% (e.g. Sample 11, 20%) increases with the shearing rate by as much as 2.1° (Fig. 4.37a). In the case of FG mixture, when the content of fine fraction is greater than 60%, increase of shearing rate leads to a

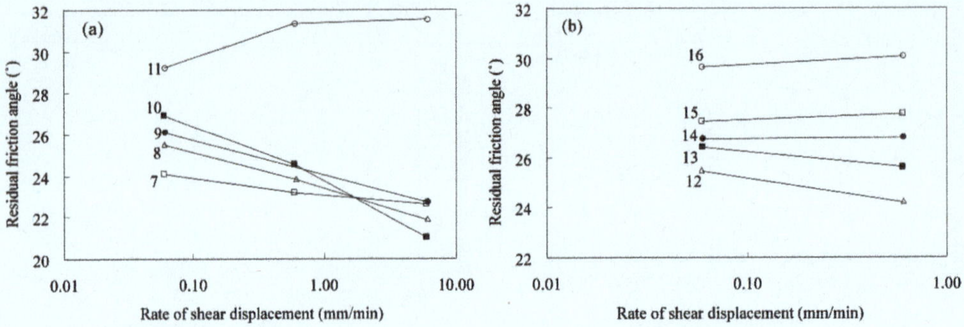


*Figure 4.37* Residual friction angle versus the shearing rate: (a) KG mixture; and (b) FG mixtures. For sample numbers, refers to Table 4.1.

decrease in the residual strength by an average of 1° (Fig. 4.37b). Residual strength of samples with fine fraction less than 60% slightly increases with shearing rate.

Based on these results, two principal types of soil behavior corresponding to shearing rate are identified. These are defined as positive and negative rate effects, depending on whether the fast residual strength stabilizes above or below the slow residual strength, respectively. Figure 4.37(a–b) indicates that the fine and/or kaolin content may be taken as an indicator for the transition of the rate effect, and the rate effects coincide well with the residual shear mode (Figs. 4.28 to 4.30). If a sample behaves in sliding mode, the fast shearing rate is to cause reduction in the residual strength. On the contrary, if a sample behaves in transitional or turbulent shear mode, the fast shearing rate is to cause increase in the residual strength.

With high fine fraction, the sample tends to form a slickensided shear surface and behaves in sliding mode. This has been confirmed by the visual inspection presented in the previous sections. The negative rate effect is considered to result from the development of pore pressures within the shear zone. The fast shearing does not allow the pore water pressure generated to dissipate. Parathiras (1994) found that if the shearing rate is gradually decreased to a value which would allow for the dissipation of pore water pressure, the shear strength of the sample gradually increases to reach that at slow shearing rate. Furthermore, the slower is the shearing rate, the less is the vertical displacement. This implies that dilation is suppressed as the shearing rate decreases.

With low fine fraction, the sample tends to behave in transitional or turbulent shear mode. At this point shearing involves interlocking of coarse particles and sliding of fine fraction. Increasing the shearing rate causes the void ratio to grow and changes the extents of interlocking and sliding by enhancing the former. Lemos (1986) reported that the fast shearing rate (above 100 mm/min) could cause a permanent disruption of the shear zone in which the fine particles have a lower degree of orientation, indicating the significance of particle interlocking.

## 4.5 CONCLUSIONS

Soils with a wide range of particle size are most commonly encountered materials in practice. A systematic investigation into the residual shear behavior of these soils was conducted by means of direct shear tests in the present study. The samples were prepared in three categories (KG, FG and FN mixtures) in terms of their composition. Based on the results of these tests, soil compressibility, the structure and water content of the shear zone and shear mode of granular and composite soils were methodically and thoroughly analyzed and discussed. Five influential factors were identified, Atterberg limits, particle size distribution, magnitude of normal stress, particle shape and shearing rate. The contributions of these factors were quantified. Five main conclusions were derived as summarized as follows.

### 4.5.1 Soil compressibility

The void ratio of the mixture, after consolidation, depends on the proportion of each particle size fraction by weight. Increasing fine fraction from zero to a threshold value, found to be approximately 20–30% in this study, causes a decrease in void ratio of the mixture. Beyond this threshold, a further increase of fine fraction causes an increase in void ratio up to the values obtained for the fine material alone.

Given the same proportion of coarse fraction (glass sand and glass beads), mixtures with only kaolin as their fine fraction (KG mixtures) have greater void ratio than those with kaolin and silt as fine fractions (FG mixtures). Synchronously increasing fine fractions of these two mixtures increases the difference between their void ratios. This suggests that particle size distribution determines maximum attainable void ratios in soils with fine fraction, and that poorer grading (as in KG mixture) results in greater void ratio.

For the same proportion of fine fractions (kaolin and silt), mixtures with artificial materials (glass sand and glass beads) as their coarse fractions (FG mixtures) develop lower void ratios than those with natural materials (river sand and crushed gravels) as coarse fractions (FN mixtures). Synchronously increasing coarse fractions of these two mixtures increases the difference between their void ratios. This indicates that the shape of coarse particles determines maximum attainable void ratios in soils with fine fraction, and that the void ratios increase with decreasing roundness or sphericity.

### 4.5.2 Water content after shearing

The differences between water contents of the shear zone and surrounding layers appear to be strongly dependent upon both the grading and magnitude of normal stress.

The high proportion of fine fraction enables the specimen to contract especially within the shear zone, where a very thin but dense layer of orientated fine particles forms, thus lowering water content of shear zone compared with surrounding layers. On the contrary, the specimens with high proportion of coarse fraction behave like dense sand with a relatively loose layer of coarse particles, thus absorbing much water.

If the specimen contains intermediate proportion of fine fraction, it tends to behave similar to dense sand when sheared under a lower normal stress (e.g. 50 kPa), thus resulting in higher water content in the shear zone compared with surrounding layers, but similar to a normally consolidated clay under a higher normal stress (e.g. 200 kPa), leading to lower water content in the shear zone.

In addition, increasing angularity of coarse particles in soils causes an increase in the relative magnitude of water content within shear zones.

### 4.5.3 Proposed shearing model for granular materials

The fluctuations of both shear stress and vertical displacement were proven to reflect the shear mechanism of granular materials. The Fast Fourier Transform (FFT) was employed successfully to decompose the measured shear stress into two components, friction and net fluctuation. It was found that the former consumes as much as 90% of the total work done by the resultant shear force while the latter consumes 10%. This indicates that though the main style of movement at residual state, from the macroscopic point of view, appears to be sliding along a wavy surface, the particle interlocking still plays an important role in the shear stress. Therefore, the averaging, which was often used to smooth the fluctuation in the literature, is not appropriate.

The magnitude of normal stress was confirmed to affect the formation of shear zone structures, which was evident from the decomposition of the components of the fluctuations by means of FFT. The higher the normal stress, the higher the degree of concentration of coarse particles, causing larger wave length and amplitude of the fluctuations.

Shearing process in granular materials can be said to take place in four stages. The first stage generally exhibits contraction due to the change of stress field caused by applied shear force. This force modifies the soil skeleton made up of the solid particles, which is shaped during the previous consolidation period. This stage ceases when granular interlocking prevents further contraction.

In the second stage, i.e. dilation before the peak, the main movement of particles occurs to overcome the interlocking leading to volume increase. This stage ceases when the particle interlocking reaches its climax, accompanied by the occurrence of the peak point. After this, the shearing enters into the third stage where the slip surface starts forming with small-size particles moving out to accommodate the larger particles penetrating the shear zone. At the end of the third stage, rigid blocks are formed and the soil reaches the residual state, like moving on a wavy surface.

### 4.5.4 Extension of the shear modes to composite soils

The residual shear modes (sliding, transitional and turbulent) put forward by Lupini *et al.* (1981) were extended for application to a wide range of soils in terms of the maximum particle size involved. The granular void ratio, $e_g$, appears to be an indicator for the transition of shear modes. The $e_g$ boundaries (0.8, 4.0) would be a useful guide to estimate the mode of residual shear behavior of natural soils with similar fine fractions and particle size ranges to those of the mixtures tested in this part of the study.

The shear mode a soil exhibits results from the counterwork of two basic processes, sliding between orientated fine particles and the interlocking of coarse particles. The orientation of fine particles favors the sliding mode whereas the particle interlocking favors the turbulent mode. Thus the resultant shear mode depends on the extents of these basic processes. When these processes operate roughly equally, the transitional mode occurs.

Though the transition of shear modes was demonstrated to be dominantly controlled by the soil gradation, normal stress and particle shape are also influential. High normal stress favors sliding mode while roughness and angularity of coarse particles intensify the particle interlocking, favoring transitional and turbulent shear modes.

### 4.5.5 Influential factors

The grain size distribution plays a pivotal role in the mobilization of residual strength. For all three mixtures involved in this study, decreasing fine fraction (<0.063 mm) from 100% to 0% resulted in a gradually transition from typical clay to granular soil behavior. When the fine fraction increased, the residual strength, generally, decreased and the residual shear behavior gradually changed from turbulent to sliding mode.

The particle shape, quantified by circularity, convexity and elongation, emerged as a significant parameter that contributes to the residual strength. Increasing circularity and/or convexity decreases residual strength while increasing elongation increases it. The roughness and angularity of coarse particles favor turbulent and transitional modes.

Atterberg limits and the magnitude of normal stress applied were confirmed to have negative influence on the residual friction angle.

The influence of normal stress becomes more significant in the lower range from 50 to 200 kPa. The higher is the normal stress, the lower is the residual friction angle. In addition, increasing normal stress favors sliding mode.

The fast shearing rate (6 mm/min compared to 0.06 mm/min) leads to a decrease in residual strength of the soils containing a sizeable fine fraction but to an increase in residual strength of the soils with a small amount of fine fraction.

The analysis of the influence of individual factors substantiates the fact that the residual strength is determined by the cooperation and/or competition of all these factors rather than any single one.

Chapter 5

# Slip zone soils of the TGP landslides

## 5.1 INTRODUCTION

Following the investigation of the shear behavior of the artificial soil mixtures presented in Chapter 4, one further step was taken in this chapter to present the shearing behavior of the natural slip zone soils in the TGP area. A total of 33 samples obtained from three landslide slip zones with different mineralogical composition and particle size distribution were tested using the large ring shear apparatus, reaching shear displacements of up to 160 cm at three different shearing rates (i.e. 0.1 mm/sec, 1.0 mm/sec and 10 mm/sec). Both residual and peak strength of these soils were examined in relation to soil properties and testing conditions. Soil properties, such as mineralogy, plastic limit, liquid limit and plasticity index, grain size distribution, which were routinely used for studying cohesive soils, and those widely used for cohesionless soils, such as particle shape, particle size and particle roughness, were all taken into account. The aim was to examine the effects of these factors on the shear behavior and to establish the shear mode of the slip zone soils in the TGP area.

A brief description of the study area and the nature of the three landslides involved are presented in Section 5.2. In Section 5.3, the testing details including the sample preparation and testing procedures is given. Testing results are presented and discussed in Section 5.4. The effects of individual influencing factors on the shear behavior are presented in Section 5.5. In Section 5.6, the fluctuations of the measured shear stress and vertical displacement were analyzed in depth. In Section 5.7, the residual shear modes, which were put forward by Lupini *et al.* (1981) for fine-grained soils and modified and extended in Chapter 4 to the composite soils, are reexamined for the ring shear testing results. The main conclusions derived from these analyses are presented in the last section.

## 5.2 STUDY AREAS AND NATURE OF INVESTIGATED LANDSLIDES

### 5.2.1 General geology

The Three Gorges was formed by deep fluvial incision of massive limestone sequences, of lower Paleozoic and Mesozoic age, along narrow fault zones, in response to Quaternary uplift (Chen *et al.*, 1995; Zhao, 1996; Li *et al.*, 2001; Liu *et al.*, 2004). As a result of this uplift and erosion cycle, Pre-Sinian metamorphic and magmatic rocks

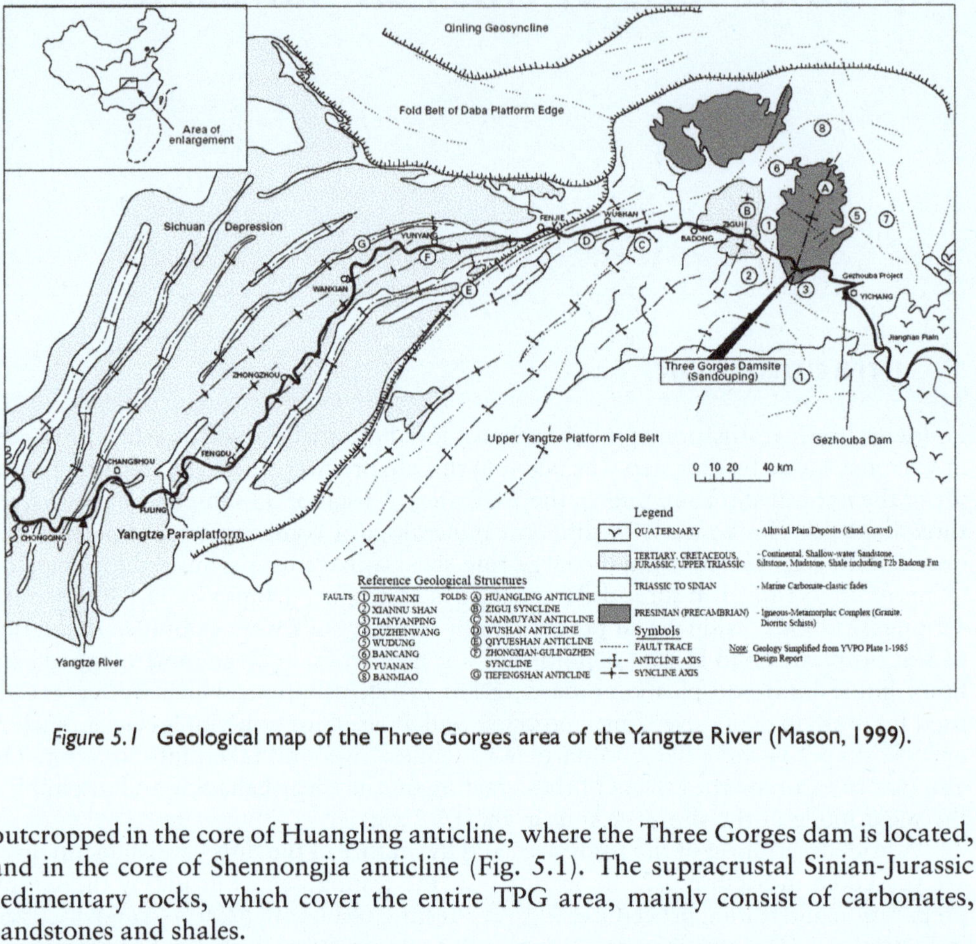


*Figure 5.1* Geological map of the Three Gorges area of the Yangtze River (Mason, 1999).

outcropped in the core of Huangling anticline, where the Three Gorges dam is located, and in the core of Shennongjia anticline (Fig. 5.1). The supracrustal Sinian-Jurassic sedimentary rocks, which cover the entire TPG area, mainly consist of carbonates, sandstones and shales.

Seismicity is weak at the Yangtze valley around the dam site, where the largest magnitude = 5.1 Zigui earthquake with an intensity of V-VI occurred in 1979 (Wu, 2001). The average rainfall in the TGP area is 100–150 mm per month while the peak monthly rainfall can be as high as 300 mm during summer months. In 1982, heavy rain caused reactivation of more than 10 landslides in Yunyang area (Xie *et al.*, 1995). In the summer of 1998, continuous rain caused more than 20 landslides in Jiangjin area, resulting in the loss of 5 lives and 500 million RMB worth of properties (Ding *et al.*, 2001; Ma *et al.*, 2005).

### 5.2.2 Nature of landslides

In the context of the present study, samples were collected from the slip zones of three landslides in the TGP area. Following is a brief description of these landslides.

The Qianjiangping landslide (QJP) occurred at the left bank of Qinggan River, a tributary of Yangtze River, on 13 July 2003 (Figs. 5.2 and 5.4). As a result, 14 people died, 10 were missing, 129 houses and 4 factories were destroyed. The tongue-shaped

*Figure 5.2* Location map of the Qianjiangping landslide (Liao *et al.*, 2005).

landslide had a volume of approximately 20 $Mm^3$, with a length of 1200 m, width of 1000 m and an average thickness of the mass displaced of approximately 40 m thinner in the upper part and thicker at the lower part (Wang *et al.*, 2004; Liao *et al.*, 2005; Li *et al.*, 2013a). The elevation of the main scarp was 450 m, and the elevation of the Qinggan River water level was 135 m when the landslide occurred. The travel distance was about 250 m. The standing trees and factory buildings on the sliding mass of the landslide indicate that the angle of the sliding surface remained constant and no rotation occurred. Field observations revealed that the slip zone consisted of a several-centimeter thick layer of silty and clayey soil with some subrounded and slickensided gravels, developed along a bedding plane at the contact between weathered mudstone and shale sequence above and relatively intact sandstone below. The location of the slip zone is shown in a simplified geological log of a borehole drilled in the middle of the landslide mass in Figure 5.3.

The slope of the mass of the Xietan landslide (XT) is some 20–30° to the horizontal as shown in Figures 5.5 and 5.6. With a length of 780 m, average width of 260 m and average thickness of 30 m, the landslide is approximately 8.9 $Mm^3$ in volume. The geologic sequence outcropping in the landslide area as shown in Figure 5.5 is briefly described in chronological order as follows (Xu *et al.*, 2003; Chai & Li, 2004): (1) Landslide deposits (Q), consisting of highly permeable (permeability approximately $0.1–8.6 \times 10^{-4}$ mm/sec) grayish-yellow well-graded gravel with little fines (GW) and rock fragments of up to 400–600 mm; (2) Middle Jurassic sediments ($J_2$), medium to thickly laminated grayish-green siltstone and quartzo-feldspathic sandstone; (3) Early Jurassic sediments ($J_1$), grayish-yellow silty mudstone and fine sandstone; (4) Late Triassic sediments ($T_3$), grayish-yellow silty mudstone and fine sandstone with shale and thin core layer and (5) Middle Triassic sediments ($T_2$), purplish-red sandstone and siltstone with mudstone. The photo in Figure 5.7, which was taken from the tunnel in Figure 5.5, indicated that the slip zone of Xietan landslide consisted mainly of a several-centimeter thick layer of grayish-yellow clayey soils with gravels. These gravels were grayish-yellow sandstone or siltstone with subangular or subrounded shape.

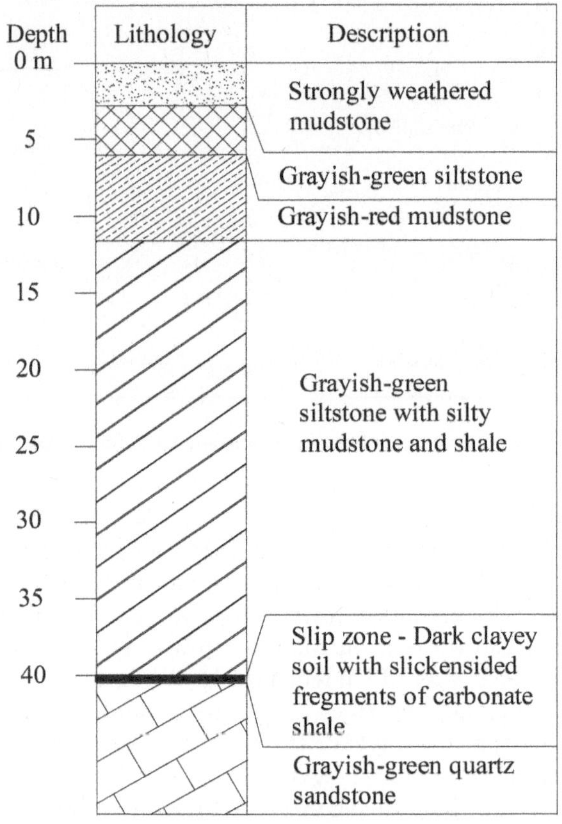


*Figure 5.3* Lithological units of the Qianjiangping landslide (Liao *et al.*, 2005).

*Figure 5.4* General view of Qianjiangping landslide (Liao *et al.*, 2005).

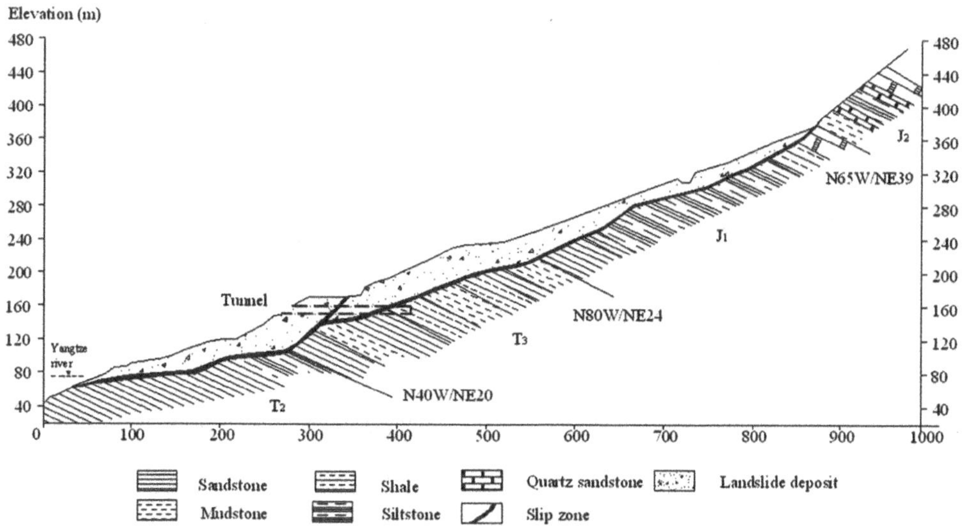


*Figure 5.5* Cross section of Xietan landslide (Xu *et al.*, 2003).

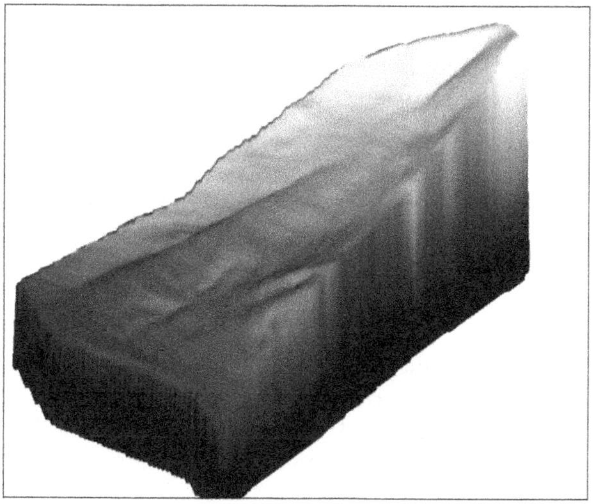

*Figure 5.6* 3D topographical model of Xietan landslide.

The Tanping landslide is located in Badong County on the left bank of Yangtze River. With its 400 m width, 450 m length and 30 m thickness, the Tanping landslide (TP) has a volume of 6.3 Mm$^3$ as shown in Figure 5.8. The sliding mass is composed of well-graded gravel-sand mixture with little cobbles and fines. Gravels and cobbles with the equivalent diameters of 10–500 mm in the mass are siltstone and silty mudstone.

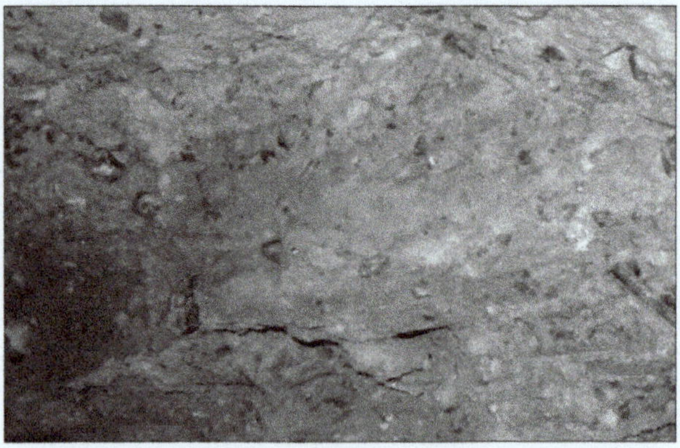

*Figure 5.7* Photo of slip zone of Xietan landslide.

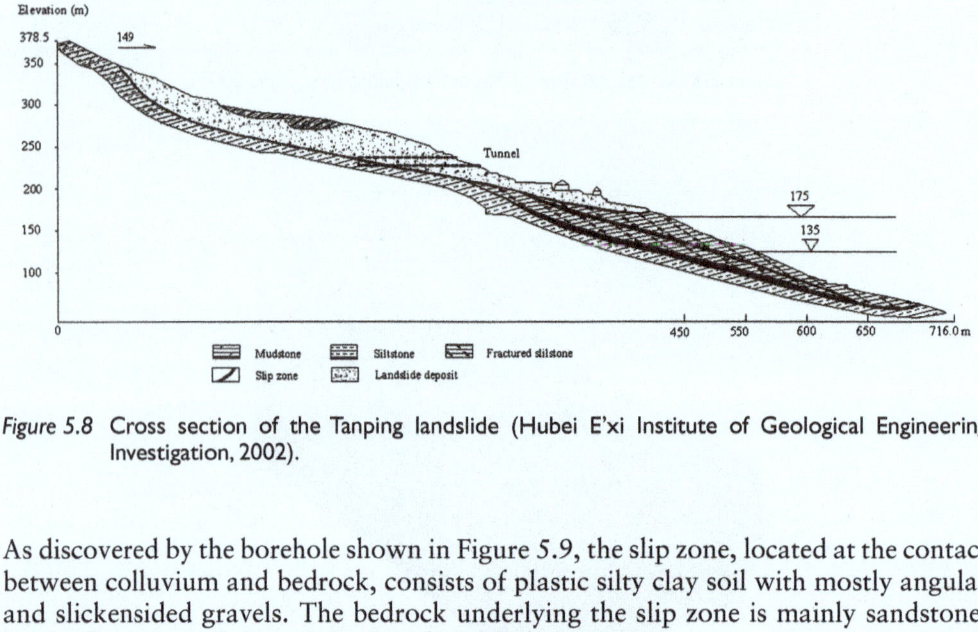


*Figure 5.8* Cross section of the Tanping landslide (Hubei E'xi Institute of Geological Engineering Investigation, 2002).

As discovered by the borehole shown in Figure 5.9, the slip zone, located at the contact between colluvium and bedrock, consists of plastic silty clay soil with mostly angular and slickensided gravels. The bedrock underlying the slip zone is mainly sandstone, marl, siltstone, and silty mudstone of Middle Triassic age.

## 5.3 EXPERIMENTAL SCHEME

### 5.3.1 Sampling and specimen preparation

For each landslide described above, three samples, each weighting about 50 kg, were obtained from the exposed shear zone in the tunnels and adits (Figs. 5.5 and 5.8). The procedures of sampling were in accordance with Chinese National Standards (CNS) GB/T50123-1999 (Standard for soil test methods) (SAC, 1999). The representative

*Figure 5.9* Slickensides on the shear surface of the Tanping landslide.

*Table 5.1* Particle size distribution of slip zone soils.

| | *Particle size distribution (weight percentage, %)* | | | | | |
|---|---|---|---|---|---|---|
| *Sample type* | *Gravel* | *Sand* | *Silt* | *Clay* | $C_u$ | $C_c$ |
| QJP1 | 18.7 | 18.6 | 43.9 | 18.8 | 26 | 0.6 |
| QJP2 | 22.2 | 17.6 | 44.8 | 15.4 | 38 | 0.5 |
| QJP3 | 20 | 18.9 | 43 | 18.1 | 100 | 0.2 |
| XT1 | 12.9 | 18.3 | 58.1 | 10.7 | 17 | 0.9 |
| XT2 | 16.7 | 17.7 | 55.4 | 10.2 | 20 | 0.8 |
| XT3 | 19.3 | 17.8 | 53.1 | 9.8 | 26 | 0.6 |
| TP1 | 16.2 | 24.8 | 46.7 | 12.3 | 69 | 0.3 |
| TP2 | 21.5 | 24.2 | 43.6 | 10.7 | 119 | 0.2 |
| TP3 | 24.2 | 23.6 | 41.1 | 11.1 | 185 | 0.1 |

undisturbed samples were well preserved and tested for obtaining the in-situ properties of the shear zone soils. Particle size distribution, Atterberg limits, dry density, void ratio, water content, particle density and saturation degree were tested in accordance with CNS GB/T50123-1999. The results of particle size distribution were listed in Table 5.1 and plotted in Figure 5.10. Other properties of share zone soils were tabulated in Table 5.2.

The classification of slip zone soils on the Casagrande's plasticity chart in terms of liquid limit and plasticity index was shown in Figure 5.11. QJP, XT and TP samples were classified as fine-grained soils with more than 50 percent of particles passing No. 200 sieve (Fig. 5.10, Table 5.1). The liquid limit and plasticity index of QJP plotted

*Table 5.2* Physical properties of slip zone soils.

| *Sample type* | *Liquid limit, $w_L$* | *Plasticity index, $I_P$* | *Particle density, $\rho_s$ (g/cm$^3$)* | *Water content, w (%)* | *Void ratio, e* | *Dry density, $\rho_d$ (g/cm$^3$)* | *Saturation degree, $S_r$ (%)* |
|---|---|---|---|---|---|---|---|
| QJP1 | 52 | 30 | 2.70 | 12.07 | 0.320 | 2.04 | 99.8 |
| QJP2 | 50 | 31 | 2.70 | 12.20 | 0.330 | 2.03 | 100 |
| QJP3 | 53 | 30 | 2.70 | 13.80 | 0.370 | 1.98 | 99.9 |
| XT1 | 31 | 11 | 2.69 | 14.90 | 0.400 | 1.92 | 100 |
| XT2 | 32 | 11 | 2.68 | 14.60 | 0.390 | 1.94 | 100 |
| XT3 | 30 | 10 | 2.69 | 14.50 | 0.390 | 1.94 | 98.7 |
| TP1 | 29 | 13 | 2.71 | 10.31 | 0.280 | 2.11 | 99.8 |
| TP2 | 27 | 12 | 2.70 | 9.30 | 0.250 | 2.16 | 100 |
| TP3 | 29 | 14 | 2.71 | 10.24 | 0.280 | 2.12 | 100 |

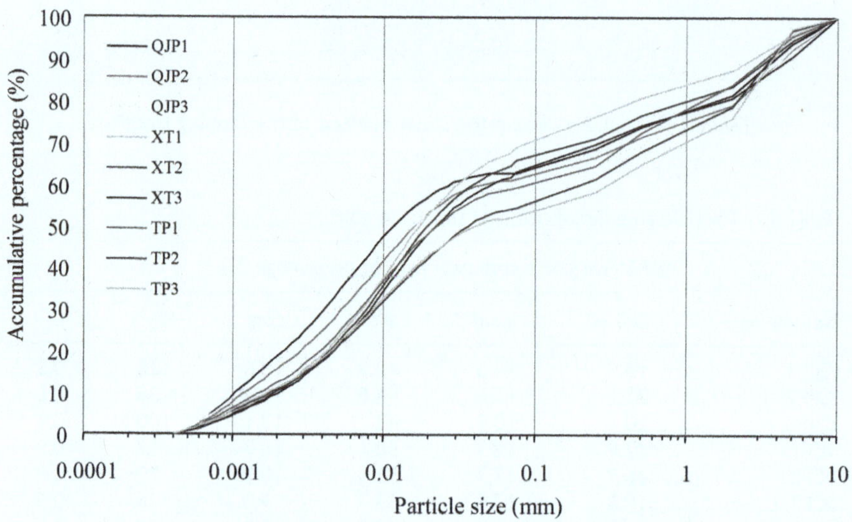


*Figure 5.10* Particle size distributions of all samples.

in CH area with its high plasticity index of 30 and liquid limit of 52 while those of XT and TP fell into CL area with moderate plasticity indexes of approximately 12 and liquid limits of approximately 30.

In this study, 33 specimens were prepared for ring shear tests (Table 5.3). The collected samples were firstly prepared by air-drying, gently mechanical crushing and sieving successively through a number of sieves to separate them into a series of particle fractions, such as clay, silt, sand and gravel. In order to reflect the true shear behavior of the slip zone soils, each specimen was prepared by taking and thoroughly mixing its constituent/s according to the particle size distribution in Table 5.1. As shown in Table 5.3, in order to test the influence of shearing rate, three duplicate specimens were prepared for each subset of slip zone soil of the Xietan and Tanping landslides. Two

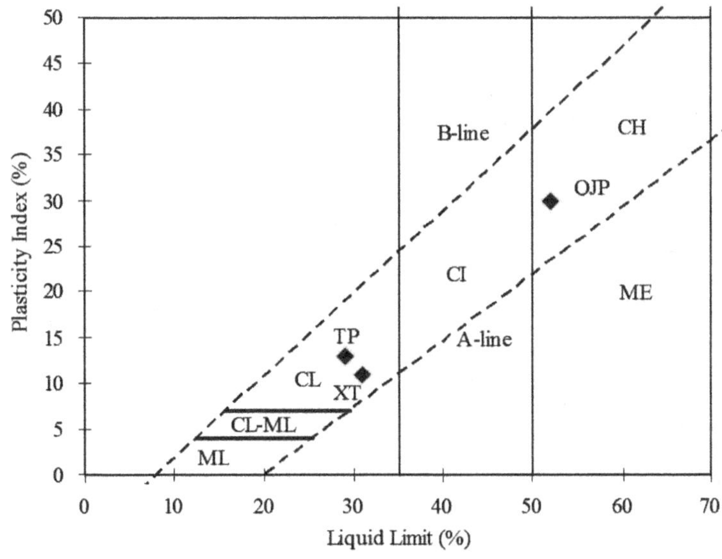


*Figure 5.11* Positions of the slip zone soils of the three landslides on the Casagrande's plasticity chart.

duplicate specimens for each subset of slip zone soil of the Qianjiangping landslide were prepared due to the limited amount of collected soils. In addition, for identifying the influence of coarse particles on the shear behavior of slip zone soils, two specimens, one of which consisted only of the coarse fraction (>0.075 mm), the other one consisted only of the fine fraction (<0.075), were also prepared for each landslide slip zone soils. The number of tested specimens and corresponding shearing rate were listed in Table 5.3. Note that the specimens denoted with suffix R were tested under the same conditions as the specimens with the same number to check the repeatability and reliability of the testing results.

### 5.3.2 Experimental setup and test procedures

In this study, a DPRI-ver.3 ring shear apparatus developed by the Disaster Prevention and Research Institute (DPRI), Kyoto University, Japan was used. Before the test program and also after every five tests, the apparatus was calibrated in accordance with the instructions of the operation manual to ensure the stability and reliability of the apparatus. The detailed description of the apparatus and specimen preparation method were presented in Chapter 3.

All tests were set up by following the same compaction procedure. Each test specimen was prepared in the sample chamber in three layers of equal thickness. Each layer was compacted by means of a tamping rod. By taking certain mass of dry soil and compacting it in a certain volume, the prepared specimen achieved the same dry density as the corresponding undisturbed slip zone soil as shown in Table 5.2. Since the saturation

*Table 5.3* Post-consolidation properties of specimens.

| *Test no.* | *Sample set* | *Water content, w (%)* | *Void ratio, e* | *Dry density, $\rho_d$ (g/cm$^3$)* | *Shearing rate (mm/sec)* |
|---|---|---|---|---|---|
| 1 | QJP1 | 11.0 | 0.30 | 2.08 | 0.1 |
| 1R* | | 10.4 | 0.28 | 2.11 | 0.1 |
| 2 | | 10.8 | 0.29 | 2.09 | 10 |
| 3 | QJP2 | 11.0 | 0.30 | 2.08 | 0.1 |
| 4 | | 10.8 | 0.29 | 2.09 | 10 |
| 5 | QJP3 | 11.7 | 0.32 | 2.05 | 0.1 |
| 6 | | 11.0 | 0.30 | 2.08 | 10 |
| 7 | XT1 | 13.7 | 0.37 | 1.97 | 0.1 |
| 7R | | 13.2 | 0.36 | 1.99 | 0.1 |
| 8 | | 14.0 | 0.38 | 1.96 | 1.0 |
| 9 | | 13.7 | 0.37 | 1.97 | 10 |
| 10 | XT2 | 13.2 | 0.36 | 1.99 | 0.1 |
| 11 | | 12.7 | 0.34 | 2.01 | 1.0 |
| 12 | | 13.7 | 0.37 | 1.97 | 10 |
| 13 | XT3 | 14.0 | 0.38 | 1.96 | 0.1 |
| 14 | | 13.2 | 0.36 | 1.99 | 1.0 |
| 15 | | 13.0 | 0.35 | 2.00 | 10 |
| 16 | TP1 | 9.0 | 0.24 | 2.17 | 0.1 |
| 16R | | 9.7 | 0.26 | 2.14 | 0.1 |
| 17 | | 8.8 | 0.24 | 2.18 | 1.0 |
| 18 | | 9.3 | 0.25 | 2.16 | 10 |
| 19 | TP2 | 8.0 | 0.22 | 2.22 | 0.1 |
| 20 | | 8.2 | 0.22 | 2.21 | 1.0 |
| 21 | | 8.6 | 0.23 | 2.19 | 10 |
| 22 | TP3 | 9.3 | 0.25 | 2.16 | 0.1 |
| 23 | | 8.6 | 0.23 | 2.19 | 1.0 |
| 24 | | 9.0 | 0.24 | 2.17 | 10 |
| 25 | QJP | 14.0 | 0.38 | 1.96 | 0.1 |
| 26 | | 13.2 | 0.36 | 1.99 | 0.1 |
| 27 | XT | 15.6 | 0.42 | 1.90 | 0.1 |
| 28 | | 14.8 | 0.40 | 1.93 | 0.1 |
| 29 | TP | 11.0 | 0.30 | 2.08 | 0.1 |
| 30 | | 10.1 | 0.27 | 2.12 | 0.1 |

degree of slip zone soil was almost 100%, the specimen then was subjected to approximately 24-hour saturation to reach fully saturation. Since the slip zones soils collected from these three landslides was subjected approximately 400 kPa normal stress in-situ, after the saturation stage, it generally took approximately 18 hours to restore this level of in situ state of consolidation. A representative plot of the consolidation is shown in Figure 5.12. Vertical deformation of the specimen was recorded during consolidation and the post-consolidation void ratio and water content were calculated as shown in Table 5.3. After completion of consolidation, the specimen was subjected to shearing at a prescribed shearing rate as listed in Table 5.3.

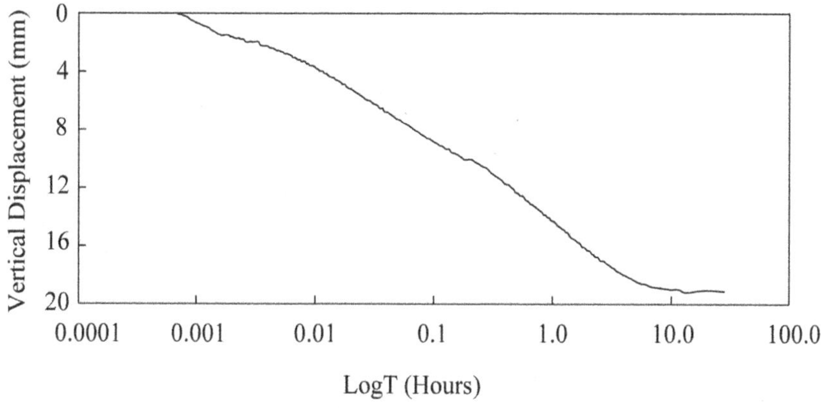


*Figure 5.12* A typical trend during the consolidation phase.

## 5.4 TESTING RESULTS

The assumption that the normal and shear stresses are distributed uniformly across the shear plane during the ring shear test is adopted as the basis for the calculation of normal stress ($\sigma$) and shear stress ($\tau$) by dividing the measured vertical load and shear force by the area of shear surface. Bishop (1971) has suggested that this oversimplification of the stress distribution on the shear plane of ring shear introduces an error which could be considered insignificant.

The test condition in this study was considered undrained as the dissipation of pore pressure can be ignored for the range of selected shearing rates (0.1 mm/sec, 1.0 mm/sec and 10 mm/sec). The effective normal stress ($\sigma'$) rather than total normal stress ($\sigma$) was used to calculate both stress ratio ($\tau/\sigma'$) and residual friction angle ($\varphi$). As described in Chapter 3, in order to prevent leakage of water and extrusion of samples during shear through the tight interface between the upper and lower rings, the DPRI ring shear apparatus had a pair of stair-shaped rubber rings with high surface smoothness placed in the specially made grooves along the perimeters of the lower inner and outer confining rings. The contact pressure between the rubber rings and the upper pair of the confining rings is adjusted at the start and maintained throughout the test at a level greater than the pore pressure that can possibly be generated in the sample and as constant as possible by means of a hydraulic piston controlled by feedback signals from a gap sensor accurate to 0.001mm. Similarly, o-rings (inserted into special slots along the inner and outer perimeters of the annular loading platen) are utilized to prevent upward leakage. High quality vacuum silicon grease is applied on to the rubber edges to lubricate and seal them. This design of DPRI ring shear apparatus made it possible to accurately measure the pore water pressure generated during shearing.

A systematic and detailed summary of various characteristic features of shear behavior of each specimen is provided in Table 5.4, including peak and residual strength (in terms of friction angle) and corresponding shear displacements at which this peak

*Table 5.4* Ring shear testing results.

| Test no. | Sample set | Strength (°) @Th. | Strength (°) @Res. | Hor. displ. @Th. (mm) | Vertical displ. (mm) @Th. | Vertical displ. (mm) @Res. | Nature of shear surface | Amplitude* (mm) Initial | Amplitude* (mm) @Res. |
|---|---|---|---|---|---|---|---|---|---|
| 1 | QJP1 | 18.2 | 13.8 | 27 | 0.1 | 1.3 | P, F | 0.015 | 0.021 |
| 1R* | | 18.2 | 13.7 | 27 | 0.1 | 0.4 | | | |
| 2 | | 20.1 | 21.6 | 27 | −0.15 | 0.2 | P, F | 0.008 | 0.012 |
| 3 | QJP2 | 19.3 | 14.2 | 30 | −0.1 | 0.7 | P, F | 0.014 | 0.022 |
| 4 | | 20.2 | 19.7 | 30 | −0.05 | 0.5 | P, F | 0.01 | 0.013 |
| 5 | QJP3 | 17.6 | 14.2 | 24 | 0.38 | 1.5 | P, F | 0.015 | 0.023 |
| 6 | | 21.3 | 22.3 | 40 | 0.22 | 0.5 | P, F | 0.009 | 0.011 |
| 7 | XT1 | 19.3 | 20.9 | 40 | 0.6 | 0.9 | U, F | 0.02 | 0.026 |
| 7R | | 18.9 | 22.0 | 40 | 1 | 2.8 | | | |
| 8 | | 19.4 | 19.8 | 32 | 1.2 | 2 | P, F | 0.015 | 0.018 |
| 9 | | 23.6 | 23.4 | 30 | 0.5 | 1.5 | P, C | 0.016 | 0.015 |
| 10 | XT2 | 18.5 | 20.3 | 28 | 0.6 | 0.9 | U, F | 0.018 | 0.024 |
| 11 | | 18.2 | 17.7 | 20 | 0.4 | 2.6 | P, F | 0.014 | 0.019 |
| 12 | | 21 | 20.4 | 24 | 0.2 | 0.8 | P, C | 0.011 | 0.017 |
| 13 | XT3 | 16.6 | 23.4 | 25 | 0.3 | 1.4 | U, F | 0.025 | 0.025 |
| 14 | | 17 | 21.9 | 30 | 0.4 | 1.5 | P, F | 0.013 | 0.019 |
| 15 | | 21.3 | 18.2 | 30 | 0.3 | 3 | P, C | 0.012 | 0.014 |
| 16 | TP1 | 21.6 | 24.9 | 50 | 0.5 | 1 | U, C | 0.042 | 0.053 |
| 16R | | 21.6 | 25.5 | 45 | 0.2 | 0.6 | | | |
| 17 | | 19.6 | 24.9 | 40 | −0.1 | 0.4 | U, C | 0.031 | 0.036 |
| 18 | | 23.1 | 25.0 | 20 | 0.25 | −0.4 | U, C | 0.018 | 0.027 |
| 19 | TP2 | 22.3 | 26.0 | 20 | 0.2 | 0.6 | U, C | 0.05 | 0.056 |
| 20 | | 21.2 | 25.0 | 20 | 0.24 | 0.6 | U, C | 0.028 | 0.041 |
| 21 | | 22.4 | 25.5 | 18 | 0.1 | 0.1 | U, C | 0.021 | 0.024 |
| 22 | TP3 | 20.6 | 25.9 | 19 | 0.1 | 0.8 | U, C | 0.034 | 0.051 |
| 23 | | 22.9 | 25.6 | 20 | 0.1 | 0.5 | U, C | 0.026 | 0.038 |
| 24 | | 23.4 | 25.1 | 21 | 0 | 0.1 | U, C | 0.015 | 0.029 |
| 25 | QJP | 11.2 | 9.4 | 18 | 0.5 | 0.8 | P | | |
| 26 | | 23.8 | 21.6 | 38 | 0 | 1.5 | U | | |
| 27 | XT | 17.6 | 16.1 | 26 | 0 | 1 | P | | |
| 28 | | 31.3 | 28.3 | 39 | 0.1 | 0.1 | U | | |
| 29 | TP | 21.1 | 18.5 | 34 | 0.1 | 0.1 | P | | |
| 30 | | 33.4 | 31.7 | 31 | 0.1 | 0.1 | U | | |

Abbreviations: @Th: at peak; @Res: at residual; P: planar; U: undulating; F: far spacing (>10 mm) between coarse particles present in the shear surface; C: close spacing (<10 mm) between coarse particles; *: amplitude of fluctuations of stress ratio.

and residual were reached, the change in specimen height at this displacement value and at the end of shearing, the nature of shear surface and the initial value of amplitude of undulation (if any) and reduction in this value when shearing was stopped.

The stress ratio ($\tau/\sigma'$) is plotted in Figure 5.13(a–c) against shear displacement and each plot is identified by test no. and shearing rate. For each set of samples, one

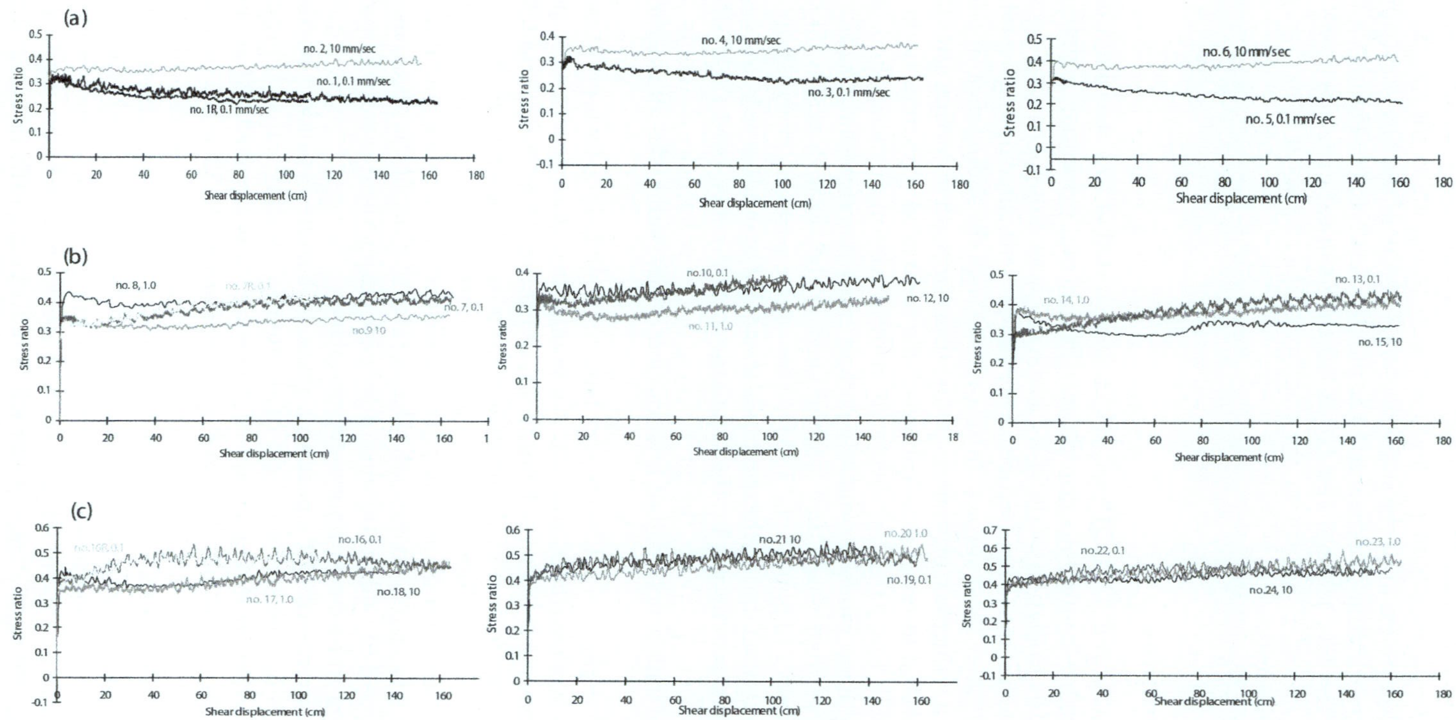


*Figure 5.13* Stress ratio versus shear displacement plots: (a) QJP set; (b) XT set; and (c) TP set.

replicate test, which is denoted by the suffix "R", was conducted, e.g. Specimen 1R has the same composition as Specimen 1 and both were tested under the identical conditions. This was done to examine the stability and reliability of the ring shear apparatus and to verify the repeatability of the tests. The plot for Test No. 1 in Figure 5.13 almost overlaps that of Test No. 1R, and Test No. 8 has the identical pattern with 8R and so do test numbers 17 and 17R, proving the stability of the ring shear apparatus and the repeatability of the tests.

### 5.4.1 QJP samples

Specimens (1, 1R, 3 and 5) of QJP set, which were tested at low shearing rate, were easily split through a continuous non-planar shear plane, presenting a sliding shear mode. The visual examination on the shear surface indicated that it had well-developed slickensides, though some coarse particles were also present, interrupting the formation of smoother and flatter shear surface. In accordance with the sliding shear mode, these samples showed a reduction (brittleness) in the shear strength occurring after the peak point. It is considered to result primarily from the formation of extremely thin layers of shear plane with fine particles orientated in the direction of shear; these particles would originally have random orientations and must therefore have had a greater resistance to shear than when they were aligned in the shear direction (Skempton, 1964; Kenney, 1977; Lupini *et al.*, 1981). The coarse particles in the shear surface would come into contact during the rotational shear. This would cause frequent particle interlocking events as reflected by the fluctuations of the measured shear stress and vertical displacement. Although, the interlocking of coarse particles cannot be avoided in such cases, the high proportion of fine fraction (more than 60%) allows the sliding mode to prevail.

Specimens 2, 4 and 6 of QJP were very hard to be split along their presumed shear planes after they were extruded from the confining rings. A view across the potential slip zone of Specimen 4 is shown in Figure 5.14, illustrating that "potential shear zone" does not possess any feature to distinguish it from its surrounding zones. This indicates the turbulent or transitional shear mode. The difficulty in splitting the specimens along their presumed shear planes was partly attributed to the rolling and interlocking of coarse particles and partly to the high shearing rate that might have prevented alignment of particles and caused continuous turbulence during shear eventually producing a highly reworked soil zone without a distinct plane.

The shear patterns of QJP specimens seem to be rate-dependent as shown in Figure 5.13a. At the low shearing rate of 0.1 mm/sec, a peak value approximately 20% lower than that at the high rate of 10 mm/sec was attained. After the peak, the low rate produced residual shear strength approximately 50% less than the high rate in terms of stress ratio. At the high rate, instead of an orderly alignment of particles, continuous turbulent mixing is considered to be responsible for the fairly constant residual strength throughout the tests, and hence significantly larger residual strength. Lemos (1986) reported that the fast shearing rate could cause a permanent disruption of the shear zone in which the fine particles have a lower degree of orientation, indicating the significance of particle interlocking in the increase of residual strength.

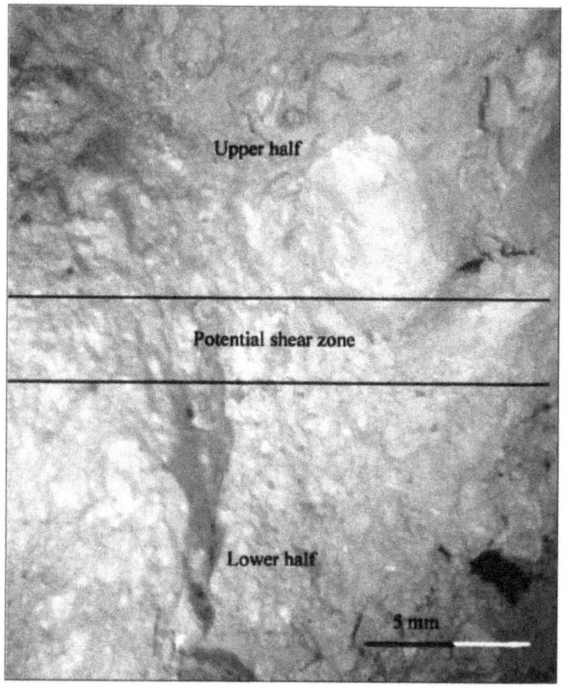


*Figure 5.14* A view across the shear zone of Specimen 4 of QJP set.

### 5.4.2 XT samples

Specimens 7, 7R, 10 and 13 of XT, which were sheared at low rates, were split along two continuous non-planar surfaces bounding the shear zone consisting of a band of dense soil with a thickness of approximately 1.0 cm along the inner and outer boundaries, gradually increasing to approximately 1.5 cm along the center line as shown in Figure 5.15. At moderate or high rate of shearing, Specimens 8, 9, 11, 12, 14 and 15 of XT developed a very thin shear zone. When split apart, the bounding surfaces had a very irregular geometry with occasional polished tracks locally disrupted by coarse particles as shown in Figure 5.16.

XT specimen set exhibited seemingly inconsistent patterns as shown in Figure 5.13b. After the specimens, tested at a low shearing rate, reached their peak strengths, they underwent shear displacement with no or little reduction in their resistance, which then increased until a shear displacement of approximately 120 cm. These specimens had the lowest clay content and highest void ratio than the other sets. The former is thought to impede particle alignment while the latter result in densification during shear, which in turn cause thickening of shear zone. In addition, presence of softer and/or brittle coarse particles may have also helped densification process. High residual stresses at low shearing rate were attributed to gradual thickening of the shear zone. At moderate and high shearing rates, the specimens repeated a similar rising

*Figure 5.15* Common view from shear surfaces of XT set sheared at slow rate.

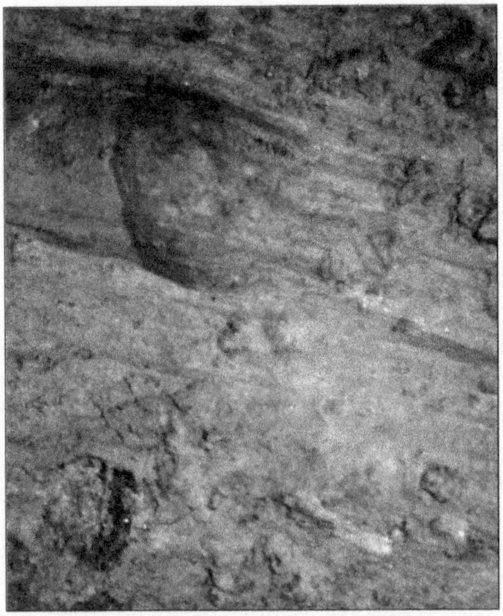

*Figure 5.16* Common view from shear surfaces of XT set sheared at moderate and high rate.

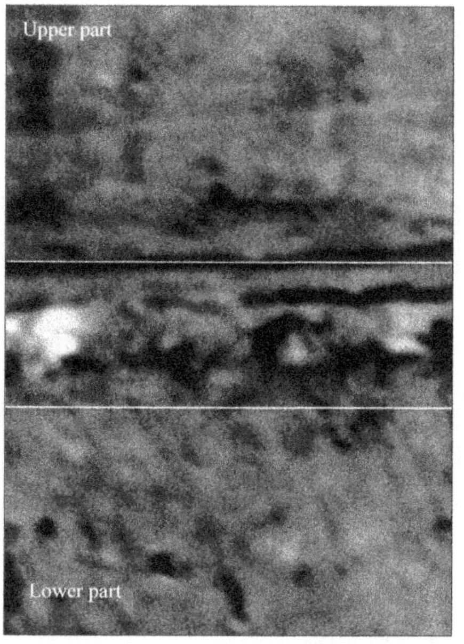


*Figure 5.17* A view across the shear zone of Specimen 16 of TP set.

pattern as those at a low shearing rate, but it appears that the densification process was not as pervasive in these samples as it also is obvious from their thin shear zones.

### 5.4.3 TP samples

The specimens in TP set (16, 16R, 17–24), regardless of the shearing rate, did not split along their presumed shear plane when they were extruded from the confining rings immediately after the tests. Visual examination of their cross-sections indicated presence of a shear zone (in between the parallel white lines on Figure 5.17 of Specimen 16) approximately 1.0 cm thick. This zone consisting of a thin band of a soft soil could be distinguished from its firm surroundings. The lower and upper boundary of the shear zone appeared to be nearly horizontal. A planar view of this shear zone after it was split with some effort is shown in Figure 5.18. The shear zone, with its discontinuous and occasional striations, appeared more like a reworked mass of soil.

The specimens of TP set (Fig. 5.13c) sheared at low shearing rate underwent an initial period of increasing resistance (within the first 30–40 cm) due to their lowest void ratio (dense nature) and largest gravel and sand fractions (compared to other samples) which inhibited formation of a single shear plane. Instead, a shear zone developed because of the difficulty of shearing through coarse particles embedded within a dense matrix. This phase was followed by a period of constant resistance and eventual reduction corresponding to particle alignment, fragmentation and/or local polishing. At high and especially moderate rate of shearing, most specimens displayed a continuously increasing trend as no alignment could take place at these faster rates.

*Figure 5.18* Plane view of the shear surface in the shear zone shown in Figure 5.17.

## 5.5 INFLUENTIAL FACTORS

Following the description in the previous section, where much emphasis was given to the shear pattern and the shear zone structure, this section evaluates, quantitatively, the relationship between the shear strength and the major influential factors, particle size distribution, Atterberg limits, particle shape and shearing rate.

### 5.5.1 Atterberg limits

The residual stress ratio is plotted against plasticity index ($I_P$) of the fine fraction (<425 μm) of the tested samples in Figure 5.19(a–b). At low shearing rate, the stress ratio decreases with the increase of $I_P$. However, at high shearing rate, no obvious relationship could be observed. The lack of correlation was also present when the residual stress ratio was plotted against either the plastic limit or liquid limit. This could be due to relatively small amount of fine fraction (approximately 50%) and low $I_P$ of the specimens. Thus, the tested soils are fundamentally different from the soils without gravels, particularly cohesive soils containing predominantly fine particles, whose residual strength is related closely to Atterberg limits, as previously reported (Skempton, 1964; Voight, 1973; Lupini *et al.*, 1981; Colotta *et al.*, 1989; Wan & Kwong, 2002; Wesley, 2003; Suzuki *et al.*, 2005; Tiwari *et al.*, 2005).

### 5.5.2 Particle size distribution

Very weak positive or negative correlations could be found between the residual stress ratio and the amount of clay and silt or gravel fractions, respectively, while the residual stress ratio appeared to be linearly proportional to sand fraction, indicating that sand size particles had a great influence on residual strength of the composite soils

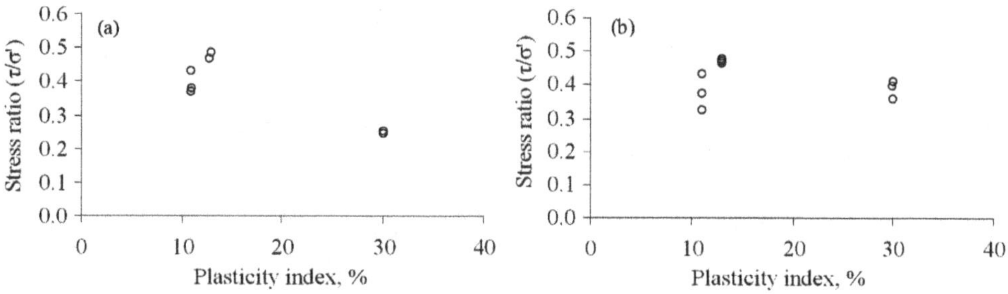


*Figure 5.19* Residual stress ratio and plasticity index: (a) at low shearing rate; and (b) at high shearing rate.

(Fig. 5.20(a–d)). In addition, residual strength appeared to increase with the ratio of gravel content to the sum of the remaining particle size fractions. Significant correlations were also observed with increasing coarse fraction (>75 μm) and the ratio of coarse fraction to fine fraction, indicating that coarse fraction played an important role in determining residual strength of these soils (Fig. 5.20(e–g)). The residual stress ratio was also closely related with the coefficient of curvature ($C_c$) as shown in Figure 5.20h. A linear correlation in semi-logarithmic space between residual stress ratio and the mean particle size is revealed in Figure 5.20i. This demonstrated again that the grain size distribution was an important factor controlling the residual strength of the soils containing coarse particles. What has been observed is in good agreement with the previous studies on this issue (Miller & Sowers, 1957; Holtz, 1960; Donaghe & Torrey, 1979; Irfan & Tang, 1992; Ng & Chui, 2003). Liu *et al.* (2006) conducted ring shear, reversal direct shear and triaxial shear tests on the disturbed (remoulded) specimens from the Xietan landslide with the coarse fraction ranging from 0% to 40%. Testing results allowed Liu *et al.* (2006) to conclude that both relative proportion and actual contents of sand and gravel strongly affect the shear strength.

### 5.5.3 Particle shape

The parameters, circularity, convexity and elongation, which are used to describe and quantify the particle shape properties, were measured by means of a stereomicroscope (Olympus SZH10) equipped with an image capture and analysis system, analySIS. Clean and dried samples were sieved into sub-fractions as done in PSD analysis. The number of randomly selected particles from each fraction was large enough to determine mean value of each population parameter with variance less than 0.01. Shape parameters for each sample were obtained by weighted sum of the mean parameter values for each fraction (i.e. scaled by their corresponding weight percentages). The images captured by this system of the gravels from the three slip zone soils are presented in Figure 5.21. The QJP samples have the greatest average value of circularity (0.77) and convexity (0.68) and the least value of the elongation (0.50), due to the roundness and smoothness of the particles; the XT and TP samples have the corresponding values of 0.73, 0.66 and 0.67 and 0.62, 0.61 and 0.88 respectively.

The relationship between these particle shape parameters and residual strength of all samples were presented in Figure 5.22 individually for each of the three shearing

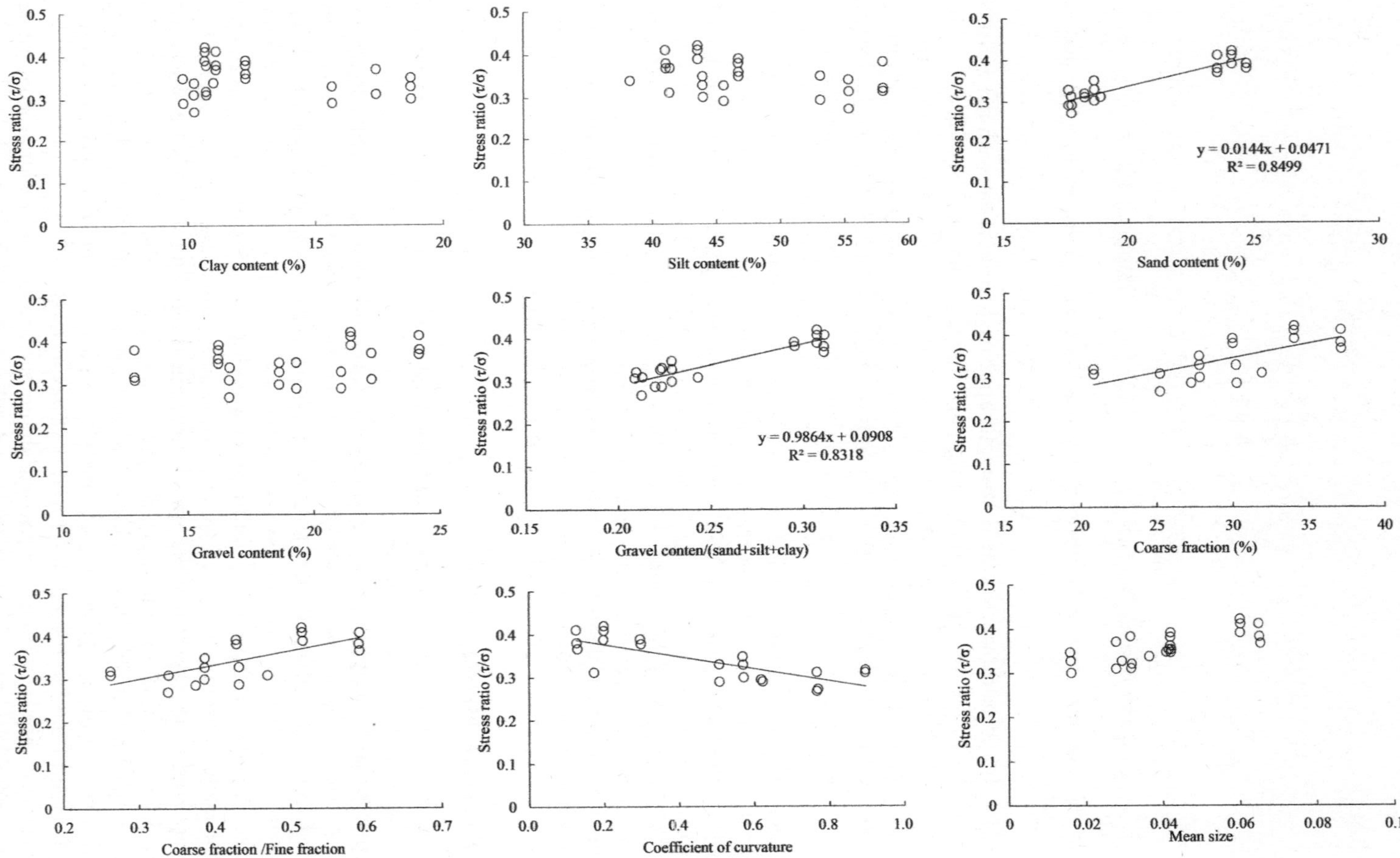


*Figure 5.20* Residual stress ratio versus various grain size fractions.

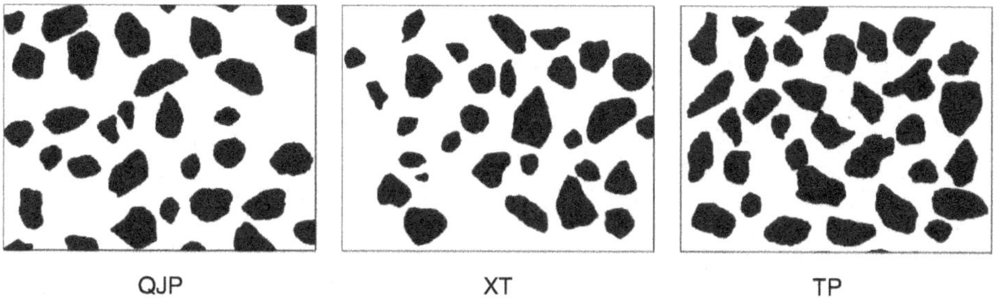


*Figure 5.21* Images for obtaining the parameters of particle shape, i.e. convexity, circularity and elongation, of different soils.

rates. Clearly, the residual strength decreases linearly with increasing circularity and convexity while increasing elongation leads to increasing of residual strength. Similar observations were made by de Graff-Johnson *et al.* (1969), Hight *et al.* (1998), Chan and Page (1997), Guimaraes (2002), Santamarina and Cho (2004) and Cho *et al.* (2006). Soil shearing involves particle rotation which prevails in loose soil, and contact slippage which prevails in dense soil. It is reasonable to postulate particle angularity hampers particle rotation while roughness holds back slippage. In the test cases of this study, the presence of elongated particles in the shear zone, if standing upward, worked as a stiff rod anchoring the upper and lower halves of sample together to prevent their relative movement.

### 5.5.4 Shearing rate

Stress ratios at peak and residual levels for fast, intermediate and slow shearing rates were plotted against the shearing rate in logarithmic scale as shown in Figure 5.23. Specimens of QJP set were sheared at two levels of shearing rates, 0.1 mm/sec and 10 mm/sec. The high shearing rate (10 mm/sec) causes an increase in both peak strength and the residual strength, exhibiting positive rate effect. The round coarse particles and the high plasticity of the fines in QJP specimens make formation of a well developed shear surface possible at lower shearing rate, and thus lower shear strength. In contrast, the high shearing rate impedes the orientation of the fine particles and leads to a relatively rougher shear surface. This again confirms the observation by Lemos (1986) that the high shearing rate could cause a permanent disruption of the shear zone in which the fine particles have a lower degree of orientation, indicating the significance of particle interlocking in the increase of residual strength.

The peak or residual shear strength of TP specimens does not seem to be influenced by shearing rates, suggesting no rate effect. This may be due to the low plasticity of the fines and the angularity of the coarse particles present in these specimens. These two properties both hinder formation of a well slickensided shear surface and offset the influence of shearing rate.

In the cases of XT specimens, there seemingly is a more complex pattern of shear strength against shearing rates. The peak strength increases with the shearing rate,

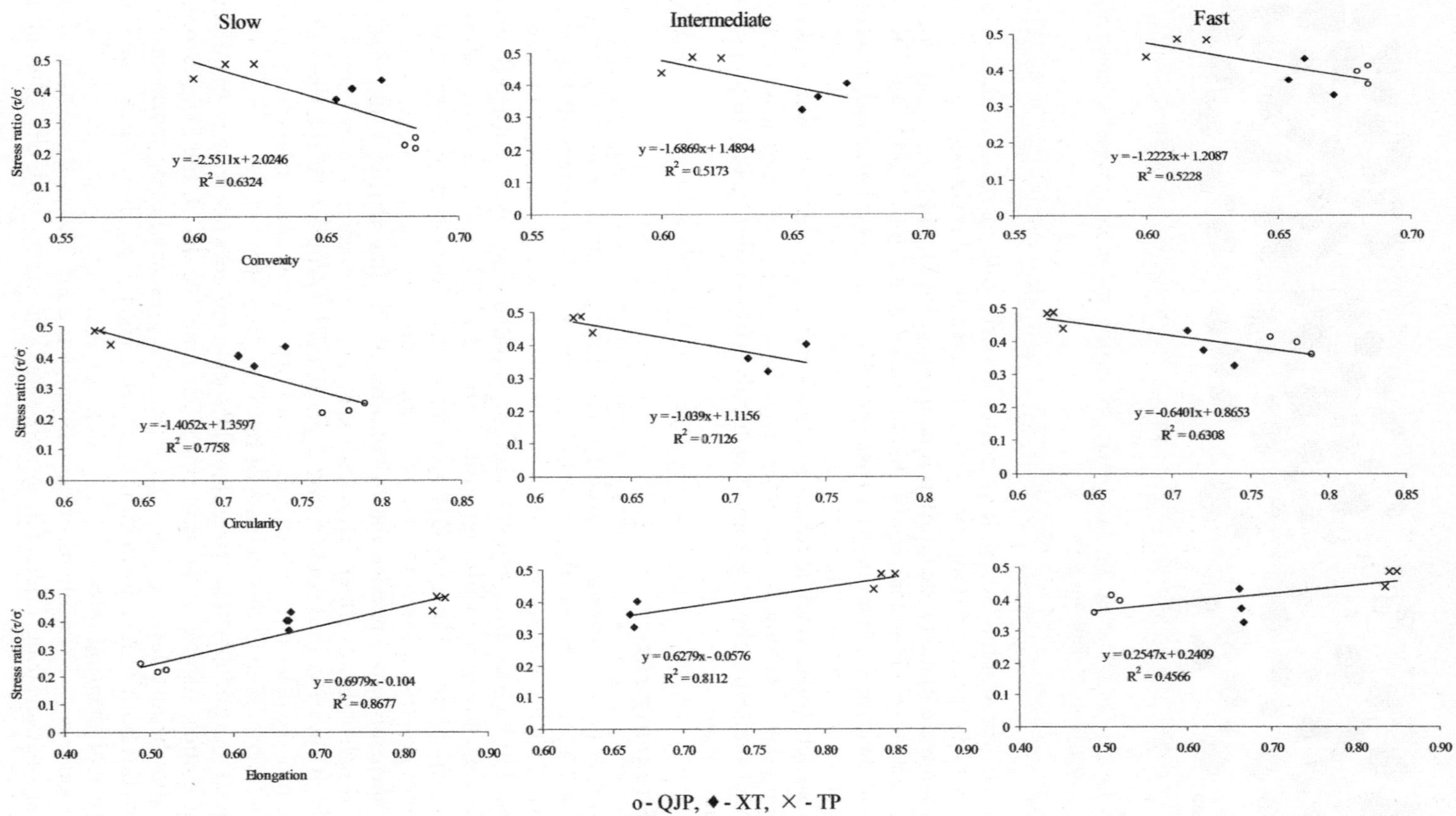


*Figure 5.22* Shear stress ratio versus particle shape parameters, i.e. convexity (first row), circularity (second row) and elongation (third row), at different shearing rates.

while the residual strength exhibits an uncertain pattern, either a concave curve with the lowest strength value at the intermediate shearing rate (1.0 mm/sec) or a monotonically decreasing behavior with increasing shearing rate. Though the exact reason for these patterns has not been clearly understood, any possible explanations should refer to the lowest clay content and highest void ratio of XT specimens than the other sets. The former is thought to impede particle alignment while the latter result in densification during shear, which in turn cause formation and expansion of a shear zone. In addition, presence of softer and/or brittle coarse particles may have also helped the densification process. High residual stresses at low shearing rate were attributed to gradual thickening of the shear zone. At moderate and high rates, though the specimens repeated a similar rising pattern, it appears that the densification process was not as pervasive which is also obvious from their thin shear zones.

## 5.6 FLUCTUATIONS OF MEASURED SHEAR STRESS

The analysis on the fluctuations of the shear stress of granular materials in Chapter 4 indicates that the Fast Fourier Transform (FFT) could be a suitable and useful method to determine the work done by the measured shear force. This work was decomposed into two, the kinetic energy ($E_k$) enabling the lateral movement and the wave energy ($E_w$) used in sustaining fluctuations. Over any period of these fluctuations, Equation 4.9 enables calculation of the percentages of the contributions of different shear modes to the shearing process as described in Chapter 4.

The fluctuations of shear stress during the ring shear tests discussed in this chapter are also pronounced (Fig. 5.13a–c). The measured data segment of the shear stress from the peak point to the end of the test was taken to analyze the fluctuations. In Figure 5.24, the solid line is the stress ratio which is directly from the measured shear stress and the dashed thick line is the main trend which is the result from 100-point average over the measured shear stress. When the red line was subtracted from the black line, the normalized fluctuation with zero as the mean value was obtained as shown in Figure 5.25. The box plot in Figure 5.26 shows the statistical characteristics of the three data sets, the measured data, the averaged "main trend" and the net fluctuation.

In statistics, box plot has its advantages to show the main properties of data in terms of the 25th, 75th, 5th and 95th percentiles and the mean value. The identical shape of box A and B indicates that the 100-point average could properly distill the fluctuation from the measured shear stress without distorting the main pattern. In the same way, the vertical displacement measured during the shearing was also decomposed into its main trend and fluctuations from this line. As done in Chapter 4, the fluctuations on both shear stress ratio and vertical displacement were in turn decomposed by means of the FFT to derive their length frequency spectrums which present the wave characteristics (length frequency (LF) and amplitude) in Figure 5.27a and b, respectively. For the fluctuations of stress ratio, the largest amplitude is located in the length frequency range from 0.1 to 0.25, indicating the main wave characteristics of the fluctuation. For the fluctuations of the vertical displacement, the largest amplitude occurs over the same range of length frequency. This indicates that the fluctuation of stress ratio is directly related to the fluctuation of the vertical displacement

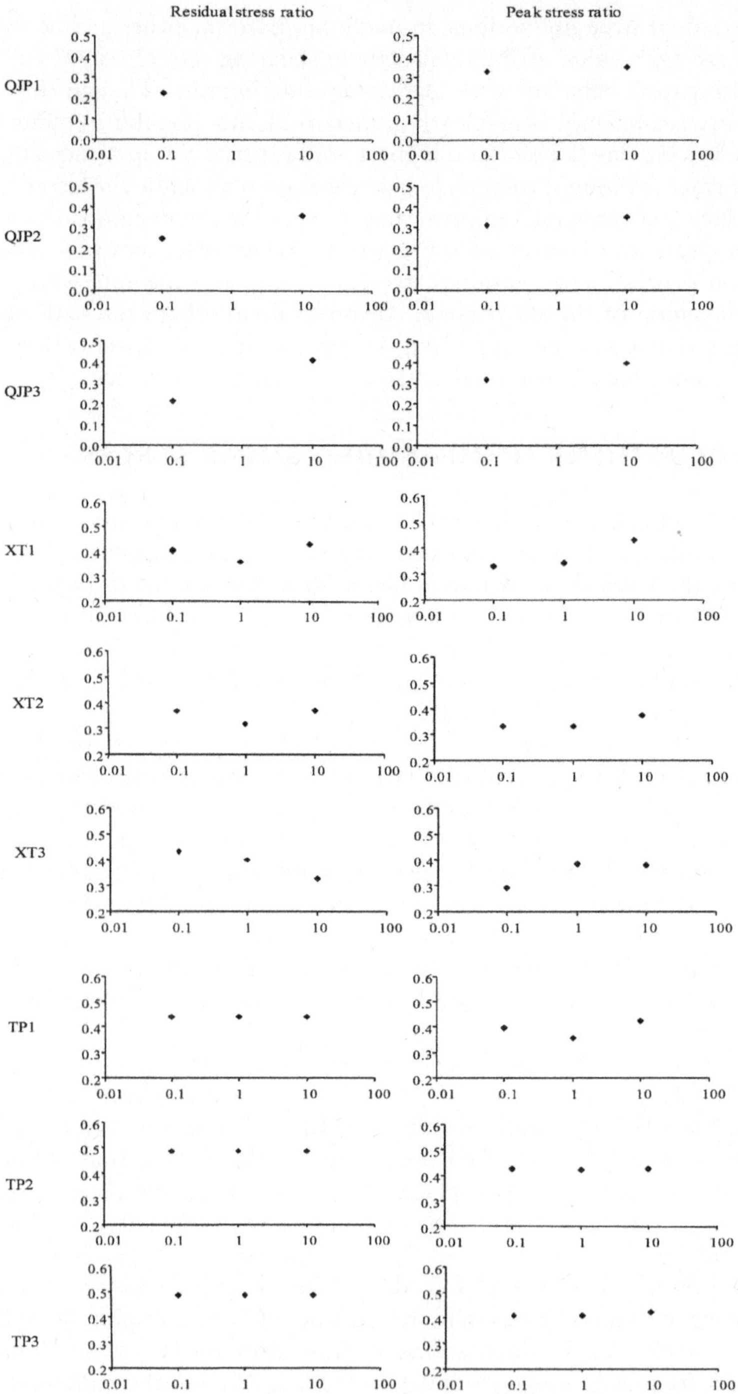


*Figure 5.23* Variations in residual (1st column) and peak (2nd column) stress ratios as a function of shearing rate for QJP, XT and TP specimen sets at three different levels of increasing gravel (or decreasing clay) contents.

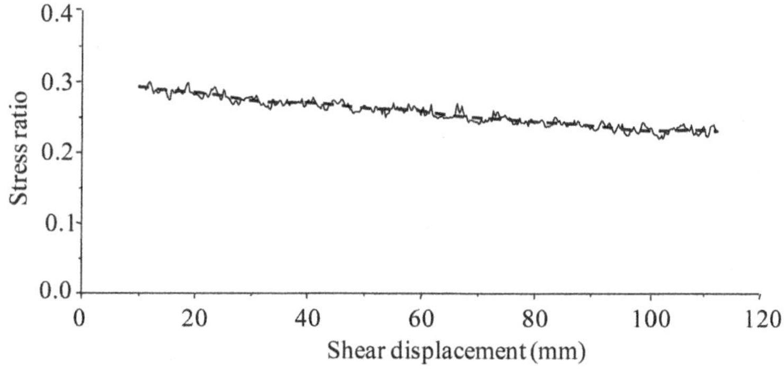


*Figure 5.24* The measured stress ratio (solid) and the 100-point average line (dashed thick) versus shear displacement of Test No. 1.

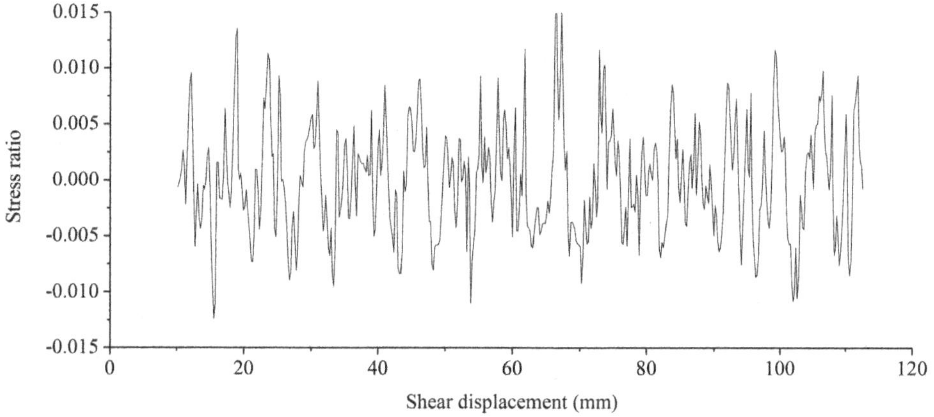


*Figure 5.25* The decomposed fluctuations of the stress ratio of Test No. 1.

and the extra energy supplied by the fluctuating shear force is mainly consumed by the fluctuation of the vertical displacement (Li & Aydin, 2013).

The main characteristics of the fluctuations of the stress ratio and the vertical displacement against the shearing rate are illustrated in Figures 5.28 to 5.30. For all three sets of specimens, the wave length and the amplitude of both the stress ratio and the vertical displacement decrease with increasing shearing rate. Due to the presence of coarse particles, the shear surface of composite soils is more likely to be a wavy surface rather than a flat one. This has been proved when the shear surface was visually examined. The coarse particles with the surrounding fines work as an obstacle and generally penetrate into the opposite block of the sample during shearing leading to the formation of deep striations on the shear surfaces (Figs. 5.15, 5.16 and 5.18). When the coarse particles in the opposite surfaces get into contact with each other, this causes a positive vertical displacement to overcome jamming and/or particle breakage, thus creating the fluctuations.

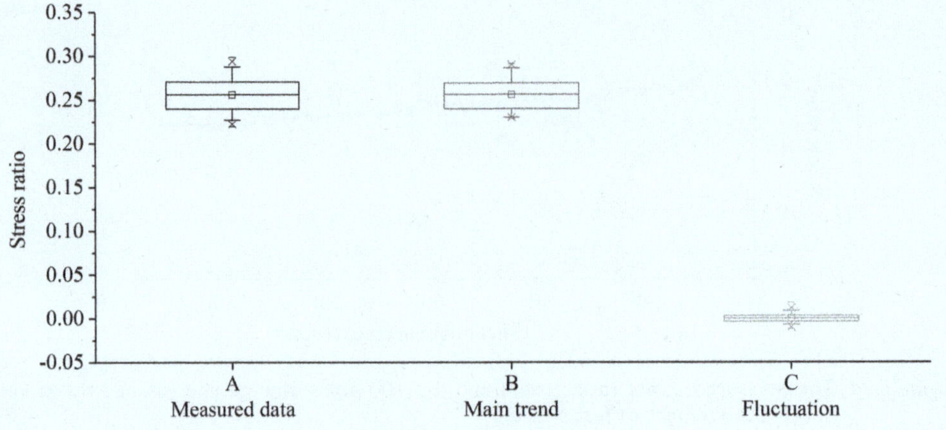


*Figure 5.26* **Box chart illustrates the loyalty of 100-point average (red) to the measured data (black) and the characteristics of the fluctuation (green).**

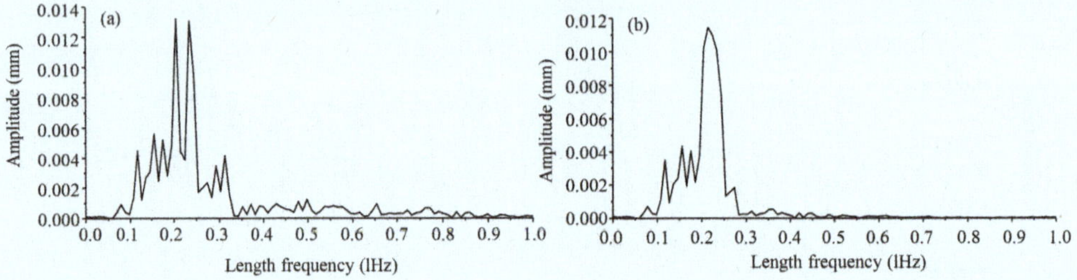


*Figure 5.27* **The length frequency spectrum of the fluctuation of Test No. 1: (a) the stress ratio; and (b) the vertical displacement.**

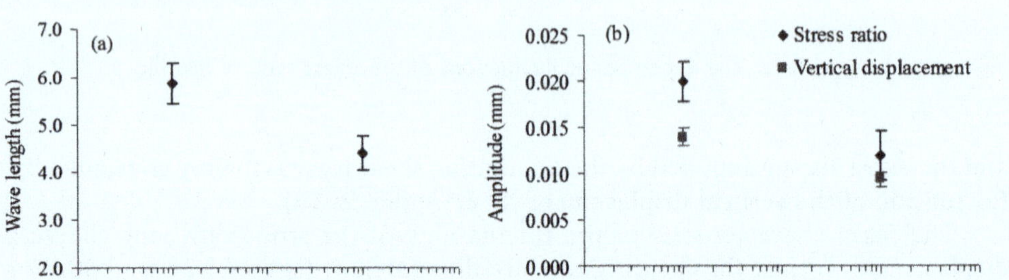


*Figure 5.28* **Fluctuation characteristic is a function of the shearing rate for QJP specimens: (a) wave length; and (b) amplitude.**

The foregoing discussion suggests that the wave length of the fluctuation of the vertical displacement reflect the irregularity of the shear surface, while the fluctuation amplitude of the stress ratio reflect the energy needed to overcome the irregularity of the shear surface. When sheared at lower shearing rate, the fines among the coarse

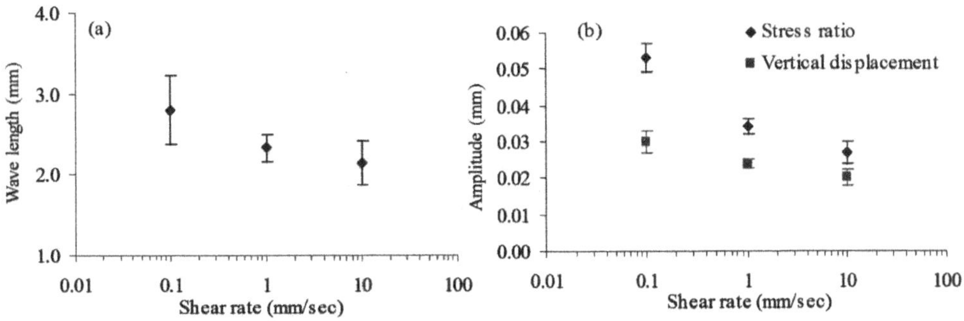


*Figure 5.29* Fluctuation characteristic is a function of the shearing rate for XT specimens: (a) wave length; and (b) amplitude.

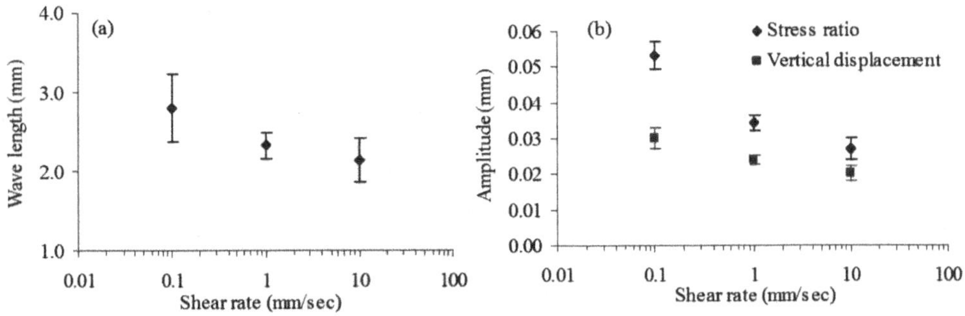


*Figure 5.30* Fluctuation characteristic is a function of the shearing rate for TP specimens: (a) wave length; and (b) amplitude.

particles were allowed to be pressed to get relatively denser and flatter, making the coarse particles to stand out. This mechanism results in increasing amplitude of the fluctuation. On the contrary, when sheared at higher rate, the fines among the coarse particles have no time to be pressed. The shear zone is more likely to be a reworked and relatively softer soil, making the amplitude of the fluctuation smaller.

As shown in Figure 5.31, among the three sets of specimens, the QJP set has the greatest wave length and smallest amplitude of the fluctuation whereas the TP set has the opposite. As shown in Figure 5.21, the specimens of QJP set contain the coarse particles with relatively smooth surface and subrounded shape, while those of TP set have the coarse particles which are rough and angular. During shearing, angularity and roughness of the coarse particles may prevent the particles from being pushed into the fine fractions and being rotated. Thus, more coarse particles are retained in the shear surface, causing shorter wave length of the fluctuations. On the other hand, frequent irregular coarse particles lead to deep striations by penetration, causing larger amplitude of the fluctuation.

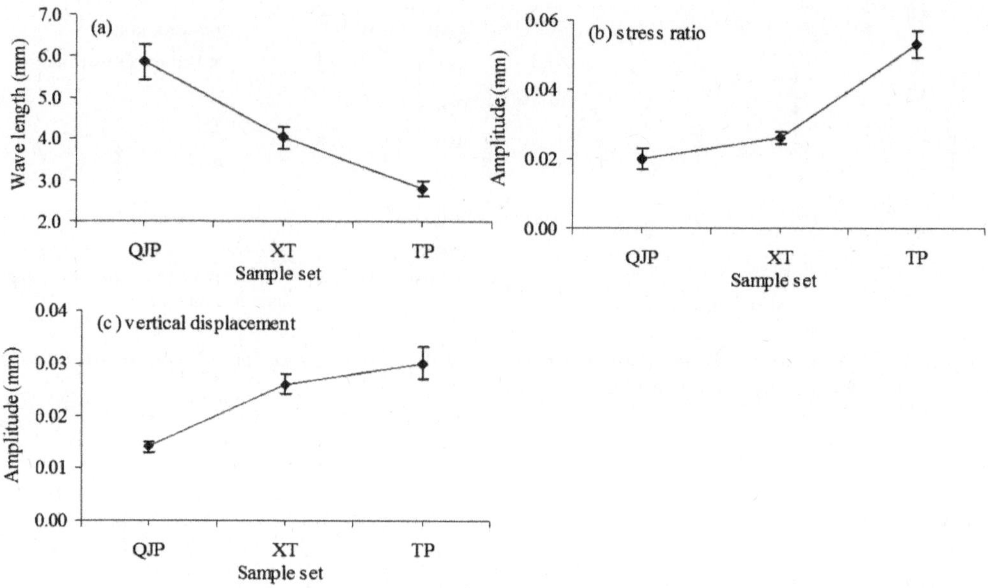


*Figure 5.31* Variations of fluctuation characteristics among the three sets of samples: (a) wave length; (b) amplitude of the stress ratio; and (c) amplitude of the vertical displacement.

## 5.7 RESIDUAL SHEAR MODES

Based on a study of the shear behavior of mixtures of sand and clay, Kenny (1977) defined the relative residual strength, $R_\varphi = [\tan\varphi'_R - (\tan\varphi'_R)_C]/[(\tan\varphi'_R)_M - (\tan\varphi'_R)_C]$, where $\varphi'_R$ = residual friction angle of the mixture; $(\varphi'_R)_C$ = residual friction angle of the clay minerals, tested on their own and $(\varphi'_R)_M$ = residual friction angle of the massive (or coarse) minerals (sand in Kenney's case) tested on their own. Lupini *et al.* (1981) related $R_\varphi$ to the granular void ratio, $e_g$ = [volume of platy particles and water]/[volume of rotund particles], to three modes of residual shear behavior of fine-grained soils, i.e. the sliding mode, the transitional mode and the turbulent mode.

The relative residual strength $R_\varphi$ was plotted against the granular void ratio $e_g$ in Figure 5.32, based on testing results of the ring shear tests of composite natural slip zone soils of TGP and the direct shear tests on artificial mixtures of clay, silt, sand and gravel presented in Chapter 4. The relationship between $R_\varphi$ and $e_g$ in Figure 5.32 reflects a transition in residual shear behavior of composite soils containing a large range of particle size. The $e_g$ range from 0.8 to 4.0 was distinguished as the transitional zone, over which the shear behavior changes from sliding to turbulent mode. All points of the composite soils are located into the transitional range. This coincides with what has been observed during the visual examination of the structures of shear zones.

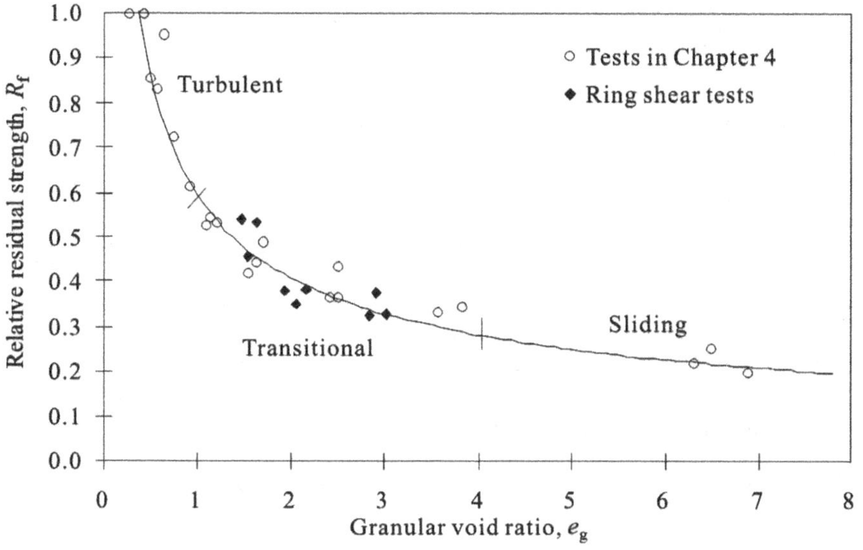


*Figure 5.32* Relative residual strength, $R_\varphi$, versus granular void ratio, $e_g$, and the boundaries of the validity domains of the shear modes.

## 5.8 CONCLUSIONS

The analyses of the experimental results reported in this chapter on composite soils collected from three slip zones of landslides in the TGP area have enabled the reaching of the following conclusions.

### 5.8.1 Influential factors

The relationship between the residual shear strength and Atterberg limits suggests, to some extent, that high plasticity index of fines in the composite soils would lower the residual shear strength. However, this correlation is not very strong since the presence of the coarse particles in such soils is to weaken the influence of Atterberg limits which can only be measured on the fine fraction and can represent the fines rather than the soil as a whole.

For soils containing abundant coarse particles, the grain size distribution plays an important role on the shear behavior. The parameters, which are derived from grain size distribution, such as the coefficient of curvature ($C_c$), the sand content, the ratio of gravel content to the sum of the remaining particles and the ratio of coarse fraction to fine fraction, show close correlations to the residual strength.

The particle shape is confirmed again as one of the main factors influencing the shear behavior. The quantitative shape parameters (circularity, convexity and elongation) measured by means of microscope and image analysis technique, correlate well with the residual strength of the composite soils. The greater the values of the circularity and convexity, the lower the residual shear strength. The greater the value of the elongation, the higher the residual strength.

The shearing rate does exert some influence on the shear behavior of composite soils. However, the effect (positive, neutral or negative) of the shearing rate is guided by the inherent properties of the soils. Particle shape, plasticity, void ratio and water content are all overriding factors compared to the shearing rate and control the shear behavior of composite soils sheared at different rates.

### 5.8.2 Fluctuations of measured shear stress

The fluctuations of both the stress ratio and the vertical displacement were analyzed with the aid of FFT. The quantified characteristics of these fluctuations (amplitude and wave length) reflect the irregularity and roughness of the shear surface. Besides normal stress, which has been testified to be a factor influencing amplitude and wave length, two factors, shearing rate and particle shape, were also demonstrated to impose some effects. Both amplitude and wave length decreases with increasing shearing rate. The angularity and roughness of coarse particles lead to high amplitude and short wave length.

### 5.8.3 Residual shear modes

The composite soils described in this chapter all showed the transitional shear mode. This was consistent with the visual examination of their shear zones. The correlation between the residual shear strength and the granular void ratio, when mapped on the relationship derived from the artificial mixtures reported in Chapter 4, show good agreements with the boundaries marking the range over which the shear mode changes from sliding to turbulent mode. These observations reinforce the validity of the framework defined by the direct shear tests on artificial mixtures in encapsulating the behavior of a large variety of natural soils regardless of the shearing method.

Chapter 6

# Conclusions

## 6.1 INTRODUCTION

It has long been recognized that the slip zones are the weakest features of ancient/old landslide masses and that the mechanical properties, especially the residual strength, of these zones play a critical role in the stability and reactivation of these former landslide masses. In the TGP area, approximately 2,490 ancient landslides of various scales have been identified during the earlier field investigations. The reactivation potential of these landslides poses a great threat to navigation in the reservoir and to operation of the TGP dam.

Effective mitigation of landslide disasters calls for systematic studies to improve our insight into mechanical properties of these slip zones. However, slip zone soils in these landslides fundamentally differ from more widely studied fine-grained and granular soils due to their pronounced properties: (1) wide range of particle size; (2) abundant coarse fraction; (3) a wide range of Atterberg limits; (4) diversified clay mineralogy; and (5) diversified properties of coarse particles. These properties hamper investigations into these composite soils due to the necessity of a shear apparatus with a large sample chamber and long shear displacement capacity, and longer testing period.

In order to improve our understanding of the shearing mechanism of such slip zone soils, a systematic laboratory testing program was conducted. A total of 138 lab tests were carried out using large ring shear apparatus (33) and intermediate direct shear box (105). Tested samples included artificial materials (glass sand and glass beads) and natural slip zone soils from the TGP landslides. The normal stress ranged from 50 to 400 kPa and the shear rate from 0.06 to 600 mm/min. Based on the results of these tests, soil compressibility, the structure and water content of the shear zone and shear mode of granular and composite soils were methodically and thoroughly analyzed and discussed. Five influential factors were identified (Atterberg limits, particle size distribution, magnitude of normal stress, particle shape and shearing rate) and the contributions of these factors were quantified. This chapter presents the major concluding remarks for the entire investigation, limitations of this study and recommendations for the future studies.

## 6.2 CONCLUSIONS

### 6.2.1 Soil compressibility

The weight percentages of each particle size fraction determine the attainable void ratios of composite soils after consolidation. Relationships between the void ratio and content of fine fraction exhibited a concave pattern. Increasing the fine fraction from zero to a threshold value, found to be approximately 20–30% in this study, caused a decrease in void ratio of the mixture. Beyond this threshold, a further increase of fine fraction caused an increase in void ratio up to the values obtained for the fine material alone.

Particle size distribution affects maximum attainable void ratios in soils with fine fraction. Poor grading (as in KG mixture) resulted in greater void ratio while good grading (as in FG mixture) resulted in lower void ratio. Given the same percentage of coarse fraction (glass sand and glass beads), mixtures with only kaolin as their fine fraction (KG mixtures) had greater void ratios than those with kaolin and silt as fine fraction (FG mixtures). Synchronously increasing fine fractions of these two mixtures increased the difference between their void ratios.

The shape of coarse particles influences maximum attainable void ratios, and the void ratios increase with decreasing roundness or sphericity. For the same percentage of fine fractions (kaolin and silt), mixtures with artificial materials (glass sand and glass beads) as their coarse fractions (FG mixtures) developed lower void ratios than those with natural materials (river sand and crushed gravels) as coarse fractions (FN mixtures). Synchronously increasing coarse fractions of these two mixtures increased the difference between their void ratios.

### 6.2.2 Water content after shearing

Whether the water content of the shear zone is greater or less than that of surrounding layers and the value (positive, zero or negative) of this difference were found to be strongly dependent upon the grading, magnitude of normal stress and particle shape.

High proportion of coarse fraction, low normal stress and angular coarse particles resulted in greater water content of shear zone than that of surrounding layers, causing positive difference between them. Low proportion of coarse fraction, high normal stress and rounded coarse particles caused the opposite.

The value of difference between water content of shear zone and surrounding layers decreased with increasing fine fraction. High proportion of fine fraction enables the specimen to contract especially within the shear zone, where a very thin but dense layer of orientated fine particles forms, thus lowering water content of shear zone compared with surrounding layers. On the contrary, the specimens with high proportion of coarse fraction behave like dense sand with a relatively loose layer of coarse particles, thus absorbing much water.

Increasing normal stress decreased the value of difference in water content of shear zone and surrounding layers. The specimens containing intermediate proportion of fine fraction tended to behave similar to dense sand when sheared under lower normal stress (e.g. 50 kPa), thus resulting in higher water content in the shear zone compared with surrounding layers, but similar to a normally consolidated clay under higher normal stress (e.g. 200 kPa), leading to lower water content in the shear zone.

Increasing angularity of coarse particles in soils caused an increase in the value of difference of water content between shear zone and surrounding layers.

### 6.2.3 Fluctuations of measured shear stress

The measured fluctuating shear stress could be decomposed into two components, friction and net fluctuation. The latter was a natural but often ignored result of the wavy and/or rough shear surface. Each of these components consumed a portion of energy supplied by resultant shear force.

It was demonstrated that the Fast Fourier Transform (FFT) can be utilized to convert the fluctuations of both shear stress and vertical displacement into signals in length frequency domain expressed in terms of wave length and amplitude, which in turn can be used to quantify the energy consumed by fluctuations.

For the artificial granular materials tested at 200 kPa, it was found that the friction consumed as much as 90% of the total work done by the resultant shear force while the net fluctuation consumed 10%. For natural slip zone soils tested at 400 kPa, the former consumed 85% and the latter consumed 15%.

The fluctuations of both stress ratio and vertical displacement were verified to coincide with structures of shear zones. The characteristics of these fluctuations (amplitude and wave length) reflected the degree of irregularity and roughness of the shear surface.

Four factors were observed to influence the fluctuations:

- Increasing fine fraction decreases the amplitude and wave length of the fluctuations;
- Increasing angularity and roughness of coarse particles increases amplitude but decreases wave length.
- Increasing normal stress increases the amplitude and wave length; and
- Increasing shearing rate decreases the amplitude and wave length.

Relationships between fluctuation characteristics and sample properties (content of fine fraction, particle shape) indicate that the fluctuation could be an indicator directly presenting inherent soil properties. It was therefore suggested that the averaging, which was often used to smooth the fluctuations in previous studies, is not appropriate.

### 6.2.4 Shear model of granular materials

The results from decomposition of fluctuations of shear stress at residual state through FFT suggested that the particle interlocking still plays an important role in the shear stress, though the main style of movement at residual state, from the macroscopic point of view, appears to be sliding along a wavy surface.

The progressive growth of wave length and amplitude of fluctuations from peak point to the onset of residual state indicated that the formation process of shear zone is accompanied by segregation (i.e. concentration of coarse particles in the shear zone and enrichment of small particles outside the shear zone).

Based on the findings of this study, a 4-stage shearing process for granular materials was proposed:

- The first stage generally exhibits contraction due to the change of stress field caused by applied shear force. This force modifies the soil skeleton made up of solid particles, which is shaped during the previous consolidation period. This stage ceases when granular interlocking prevents further contraction.
- In the second stage, i.e. dilation before the peak, movement of particles occurs to overcome the interlocking leading to volume increase. This stage ceases when the particle interlocking reaches its climax, accompanied by the occurrence of the peak point.
- During the third stage, the slip surface starts forming with small-size particles moving out to accommodate larger particles penetrating the shear zone.
- At the end of the third stage, the upper and lower halves of the sample behave as rigid blocks and the soil gradually reaches the residual state.

### 6.2.5 Shear mode of composite soils

The residual shear modes (sliding, transitional and turbulent) put forward by Lupini *et al.* (1981) were extended for application to a wider range of soils in terms of the maximum particle size involved.

The granular void ratio, $e_g$, appears to be a good indicator for the transition of shear modes. The $e_g$ boundaries (0.8, 4.0), marking the range over which the shear mode changes from sliding to turbulent, was shown to be a useful guide to estimate the residual shear mode of natural soils with similar fine fractions and particle size ranges to those of the mixtures tested in this study.

The shear mode of a composite soil results from the counterwork of two basic processes, sliding between orientated fine particles and the interlocking of coarse particles. The orientation of fine particles favors the sliding mode whereas the particle interlocking favors the turbulent mode. Thus the resultant shear mode depends on the extents of these basic processes. When these processes operate roughly equally, the transitional mode prevails.

Besides grading, which has been demonstrated to dominantly control the transition of shear modes, normal stress and particle shape were also found to impose some effects on this transition. High normal stress favors sliding mode while roughness and angularity of coarse particles intensify the particle interlocking, favoring transitional and turbulent shear modes.

### 6.2.6 Influential factors

The particle size distribution was again verified to play a pivotal role in the mobilization of residual strength. For all samples involved in this study, increasing fine fraction (<0.063 mm) from 0% to 100% resulted in a gradual transition from typical clay to granular soil behavior. When the fine fraction increased, the residual strength generally decreased and the residual shear behavior gradually changed from turbulent to sliding mode.

The particle shape, quantified by circularity, convexity and elongation, emerged as a significant parameter that contributes to the residual strength. Increasing circularity and/or convexity decreases residual strength while increasing elongation increases it. The roughness and angularity of coarse particles favor turbulent and transitional modes.

The Atterberg limits and magnitude of normal stress were confirmed to have negative influence on the residual shear strength. The influence of normal stress was found to become more significant in the lower range from 50 to 200 kPa. The higher the normal stress, the lower the residual shear strength. In addition, increasing the normal stress favors sliding mode. The fast shearing rate (6 mm/min compared to 0.06 mm/min) was demonstrated to result in a decrease in residual strength of the soils containing a sizeable fine fraction (about >50%) but in an increase in residual strength of the soils with a small amount (about >40%) of fine fraction.

The analysis of the influence of individual factors substantiates the fact that the residual strength is determined by the cooperation and/or competition of all these factors rather than any single one.

## 6.3 LIMITATIONS

Two-dimensional measurement of particle shape by using analySIS may cause some deviations from the true three-dimensional nature of soil particles.

Since the maximum particle size was 10 mm in this study, the derived conclusions may only represent the shear behavior of natural soils with similar fine fractions and particle size ranges to those of the samples tested in this study.

Although the observations from this study confirmed that there exists no significant influence of test methods on shear behavior of composite soils, the validity of this observation for other test methods, such as triaxial compression, hollow cylinder and torsional shear needs to be investigated.

## 6.4 RECOMMENDATIONS FOR FUTURE STUDY

On the basis of the present investigation, a number of research topics have emerged as of potential value for future investigations:

- Extending the tests to cover a wider variety of natural slip zone soils in terms of maximum particle size, varying particle shapes and clay mineralogy to generalize the patterns of residual shear behavior observed in this study.
- Further evaluation of the fluctuation of shear stress and vertical displacement as a general index delineating inherent properties of soils and testing conditions.
- Modifications of proposed shearing model for granular materials by means of numerical simulations to enhance the validity and reliability of experimental findings.
- Comparative evaluation of shear behavior of composite soils tested with different types of apparatuses.

# References

Allen, T. (1990) *Particle Size Measurement.* 4th edition. London, Chapman & Hall.

Alshibli, K.A. & Alsaleh, M.I. (2004) Characterizing surface roughness and shape of sands using digital microscopy. *Journal of Computing in Civil Engineering*, 18 (1), 36–45.

Bareither, C.A., Benson, C.H. & Edil, T.B. (2008) Reproducibility of direct shear tests conducted on granular backfill materials. *Geotechnical Testing Journal*, 31 (1), 1–11.

Barrett, P.J. (1980) The shape of rock particles: A critical review. *Sedimentology*, 27, 291–303.

Bernander, S., Svensk, I., Holmberg, G., Bernander, J. & Isacsson, K. (1985) Shear strength and deformation properties of clays in direct shear tests at high strain rates. In: *Proc 11th Int Conf on Soil Mech Foud Eng, 12–16 Aug 1985, San Francisco, USA.* Rotterdam, Balkema, A.A. pp. 987–990.

Bishop, A.W. (1967) Progressive failure – With special reference to the mechanism causing it. In: *Proc Geotech Conf on Shear Strength P Nat soils Rocks, 19–22 Sept 1967, Oslo, Norway*, Vol. 2. Oslo, Norwegian Geotechnical Institute. pp. 142–150.

Bishop, A.W. (1971) Shear strength parameters for undisturbed and remoulded specimens. In: Parry, R.H.G., Foulis, G.T. & Ltd. Co. (eds.) *Proc Roscoe Memorial Symp, 29–31 Mar 1971, Cambridge, UK.* pp. 3–58.

Bishop, A.W., Webb, D.L. & Lewin, P.I. (1965) Undisturbed samples of London clay from the Ashford common shaft: Strength–effective stress relationships. *Geotechnique*, 15 (1), 1–31.

Bishop, A.W., Green, G.E., Garga, V.K., Andresen, A. & Brown, J.D. (1971) New ring shear apparatus and its application to the measurement or residual strength. *Geotechnique*, 21 (4), 273–328.

Blondeau, F. (1973) The residual shear strength of some French clays: Measurement and application to a natural slope landslide. *Geological Application*, 8 (1), 125–141.

Blondeau, F. & Josseaume, H. (1976) *Mesure de la résistance résiduelle en laboratoire. Rapport du Département des sols des fondations.* Laboratoire Central des Ponts et Chaussées.

Borowicka, H. (1965) The influence of the colloidal content on the shear strength of clay. In: *Proc 6th Int Conf on Soil Mech Foud Eng, 15–18 Oct 1965, Montreal, Canada*, Vol. 1. Toronto, ON, University of Toronto Press. pp. 175–178.

Bowman, E.T., Soga, K. & Drummnond, W. (2001) Particle shape characterization using Fourier descriptor analysis. *Geotechnique*, 51 (6), 545–554.

Brabb, E.E. & Harrod, B.L. (1989) Landslides: Extent and Economic Significance. In: Brabb, E.E. & Harrod, B.L. (eds.) *Proc 28th Int Geol Congr: Symp on Landslides, Eurock 1989, 17 July 1989, Washington DC, USA.* Rotterdam, Balkema, A.A. pp. 38–50.

Brandon, T.L., Rose, A.T. & Duncan, J.M. (2006) Drained and undrained strength interpretation for low-plasticity silts. *Journal of Geotechnical and Geoenvironmental Engineering*, 132 (2), 250–257.

British Standards Institution (1990a) British Standard 1377. London, BSI.

British Standards Institution (1990b) British Soil Classification Standard (BSCS). London, BSI.

Bromhead, E.N. (1979) A simple ring shear apparatus. *Ground Engineering*, 12 (5), 40–44.

Bromhead, E.N. (1980) Discussion on "Large landslides in London clay at Herne Bay kent". *Quarterly Journal of Engineering Geology and Hydrogeology*, 13, 63–65.

Bucher, F. (1975) *Die Restscherffestigkeit natürlicher Böden. Ihre Einflussgrössen und Beziehungen als Ergebnis experimenteller Untersuchungen.* Institut für Grundbau und Bodenmechanik Eidgenössische Technische Hochschule. Report Number: 103.

Cancelli, A. (1977) Residual shear strength and stability analysis of a landslide in fissured overconsolidated clays. *Bulletin of the International Association of Engineering Geology*, 16, 193–197.

Casagrenda, A. (1948) Classification and identification of soils. *Transactions of the American Society of Civil Engineers*, 8, 901–930.

Chai, H.J., Dong, Y. & Chen, Q.Y. (2004) Overview of soil-stone high embankment construction study. *Rock and Soil Mechanics*, 25 (6), 1005–1010.

Chai, J.R. & Li, S.Y. (2004) Coupling analysis of seepage and stress fields in Xietan landslide in Three Gorges Region (in Chinese). *Chinese Journal of Rock Mechanics and Engineering*, 23 (8), 1280–1284.

Chan, L.C.Y. & Page, N.W. (1997) Particle fractal and load effects on internal friction in powders. *Powder Technology*, 90, 259–266.

Chandler, R.J. (1966) The measurement of residual strength in triaxial compression. *Geotechnique*, 16 (3), 181–196.

Chandler, R.J. (1969) The effect of weathering on the shear strength properties of Keuper marl. *Geotechnique*, 19 (3), 321–334.

Chandler, R.J. (1977) Back analysis techniques for slopes stabilization works: A case record. *Geotechnique*, 27 (4), 479–495.

Chattopadhyay, P.K. (1972) *Residual Shear Strength of Some Pure Clay Minerals*. PhD Thesis. University of Alberta, Edmonton, Canada.

Chen, Q., Hu, H., Sun, Y. & Tan, C. (1995) Assessment of regional crustal stability and its application to engineering geology in China. *Episodes*, 18, 69–72.

Cho, G.C., Dodds, J. & Santamarina, J.C. (2006) Particle shape effects on packing density, stiffness, and strength: Natural and crushed sands. *Journal of Geotechnical and Geoenvironmental Engineering*, 132, 591–602.

Clark, N.N. (1987) A new scheme for particle shape characterization based on fractal harmonics and fractal dimensions. *Powder Technology*, 51, 243–249.

Clausen, J.A. & Gabrielsen, R.H. (2002) Parameters that control the development of clay smear at low stress states: An experimental study using ring-shear apparatus. *Journal of Structural Geology*, 24, 1569–1586.

Collin, A. (1846) *Landslides in Clays* (W.R. Schriever, Trans.). Toronto, ON, University of Toronto Press.

Colotta, T., Cantoni, R., Pavesi, U., Roberl, E. & Moretti, P.C. (1989) A correlation between residual friction angle, gradation and the index properties of cohesive soils. *Geotechnique*, 39 (2), 343–346.

Cornforth, D.H. (2005) *Landslides in Practice Investigation, Analysis, and Remedial/Preventative Options in Soils*. New York, NY, John Wiley & Sons.

Coulomb, C.A. (1773) Essai sur une application des regles des maximis et minimis a quelques problemes de statique relatifs a l'architecture. *Memoires Presentes Par Divers Savants*, 7, 343–382.

Cruden, D.M. & Varnes, D.J. (1996) Landslide types and processes. In: Turner, A.K. & Schuster, R.L. (eds.) *Eurock 1996: Landslides Investigation and Mitigation. National Research*

*Council, Transportation Research Board*. Washington, DC, National Academy Press. pp. 36–75.

Cubrinovski, M. & Ishihara, K. (2002) Maximum and minimum void ratio characteristics of sands. *Soils and Foundations*, 42 (6), 65–78.

Cullen, R.M. & Donald, I.B. (1971) Residual strength determination in direct shear. In: *Proc 1st Aust–NZ Conf Geomech, Melbourne, Australia*, Vol. 1. pp. 1–10.

Dai, F.C., Deng, J.H., Tham, L.G., Law, K.T. & Lee, C.F. (2004) A large landslide in Zigui county, Three Gorges Area. *Canadian Geotechnical Journal*, 41, 1233–1240.

Day, R.W. (1989) Relative compaction of fills having oversized particles. *Journal of Geotechnical Engineering – ASCE*, 115, 1487–1491.

De Beer, E. (1967) Shear strength characteristics of the "Boom Clay". In: *Proc Geotech Conf on Shear Strength P Nat soils Rocks, 19–22 Sept 1967, Oslo, Norway*, Vol. 1. Oslo, Norwegian Geotechnical Institute. pp. 83–88.

De Graft-Johnson, J.W.S., Bhatia, H.S. & Gidigasu, D.M. (1969) The strength characteristics of residual micaceous soils and their application to stability problems. In: *Proc 7th Int Conf on Soil Mech Found Eng, Mexico*. pp. 165–172.

Dewoolkar, M.M. & Huzjak, R.J. (2005) Drained residual shear strength of some claystones from Front Range. *Journal of Geotechnical and Geoenvironmental Engineering*, 131 (12), 1543–1551.

Dikau, R., Brunsden, D., Schrott, L. & Isen, M.-L. (1996) *Landslide Recognition, Identification, Movement and Courses: Report of the European Commission Environment Programme*. London, John Wiley & Sons.

Ding, R.J., Wang, C.M. & Hu, Z.H. (2001) Seismic and geological disasters in Three Gorges Reservoir area and their prevention and control (in Chinese). *Journal of Natural Disasters*, 10 (2), 71–78.

Donaghe, R.T. & Torrey, V.H. (1979) Scalping and replacement effects on strength parameters of earth–rock mixtures. In: *Proc 7th Eur Conf on Soil Mech Found Eng, Sep 1979, Brighton, UK*, Vol. 2. London, British Geotechnical Society. pp. 29–34.

Dyskin, A.V., Estrin, Y., Kanel-Belov, A.J. & Pasternak, E. (2001) Toughening by fragmentation – How topology helps. *Advanced Engineering Materials*, 3 (1), 885–888.

Early, K.B. & Skempton, A.W. (1972) Investigations of the landslide at Walton's Wood, Staffordshire. *Quarterly Journal of Engineering Geology and Hydrogeology*, 5, 19–41.

Fleischer, S. & Scheffler, H. (1979) Problem-oriented shearing technologies for the determination of the drained shear strength of cohesive soils. In: *Proc 7th Eur Conf on Soil Mech Found Eng, Brigthon, UK*, Vol. 2. pp. 41–48.

Folk, R.L. (1955) Student operator error in determination of roundness, sphericity and grain size. *Journal of Sedimentary Research*, 25 (4), 297–301.

Fragaszy, R.J., Su, W., Siddiqi, F.H. & Ho, C.L. (1992) Modeling strength of sandy gravel. *Journal of Geotechnical Engineering – ASCE*, 118, 920–935.

Fukuoka, H., Sassa, K., Wang, G. & Sasaki, R. (2006) Observation of shear zone development in ring-shear apparatus with a transparent shear box. *Landslides*, 3 (2), 239–251.

Fukuoka, H., Sassa, K. & Wang, G. (2007) Influence of shear speed and normal stress on the shear behavior and shear zone structure of granular materials in naturally drained ring shear tests. *Landslides*, 4, 63–74.

Garga, V.K. (1970) *Residual Shear Strength Under Large Strains and the Effect of Sample Size on the Consolidation of Fissured Clay*. PhD Thesis. London, University of London.

Georgiannou, V.N., Burland, J.B. & Hight, D.W. (1990) The undrained behaviour of clayey sands in triaxial compression and extension. *Geotechnique*, 40 (3), 431–450.

Gibson, R.E. & Henkel, D.J. (1954) Influence of duration of tests at constant rate of strain on measured "drained" strength. *Geotechnique*, 4 (1), 6–15.

Gruner, H.E. & Haefeli, R. (1934) Beitrag zue untersuchung des physikalischen und statischen verhaltens koharenter boden. *Schweizer, Bauzeitung*, 103, 171–174.

Guimaraes, M. (2002) *Crushed Stone Fines and Ion Removal from Clay Slurries–Fundamental Studies*. PhD Thesis. Atlanta, GA, Georgia Institute of Technology.

Haefeli, R. (1951) Investigation and measurement of the shear strength of saturated cohesive soils. *Geotechnique*, 2 (3), 186–207.

Herrmann, H.G. & Wolfskill, L.A. (1966) Residual shear strength of weak shales. *Technical Engineering Properties of Nuclear Craters*. Cambridge, MA, Massachusetts Institute of Technology. Report number: 3-699.

Hight, D.W., Georgiannou, V.N., Martin, P.L. & Mundegar, A.K. (1998) Flow slides in micaceous sand. In: Yanagisawa, E., Moroto, N. & Mitachi, T. (eds.) *Eurock 1998: Proc Int Symp on Probl Soils, Sendai, Japan*. Rotterdam, Balkema, A.A. pp. 945–958.

Holtz, W.G. (1960) Discussion to session "Testing equipment, techniques, and errors". In: Turnbull, W.J., Hvorslev, M.J., Hilf, J.W., Barron, R.A., Peck, R.B. & Seed, H.B. (eds.) *Proc Am Soc of Civil Eng Res Conf on Shear Strength of Cohesive Soils, Colorado, USA*. Boulder, CO, American Society of Civil Engineers, University of Colorado. pp. 997–1002.

Holtz, W.G. & Gibbs, H.J. (1956) Triaxial shear tests on previous gravelly soils. *Journal of the Soil Mechanics and Foundations Division – ASCE*, 82 (SM1), 1–22.

Holtz, W.G. & Willard, M. (1956) Triaxial shear characteristics of clayey gravel soils. *Journal of Soil Mechanics and Foundation Engineering*, 82, 143–149.

Horn, H.M. & Deere, D.U. (1962) Frictional characteristics of minerals. *Geotechnique*, 12 (4), 319–335.

Hsu, T.S. & Saxena, S.K. (1991) A general formula for determining density of compacted soils with oversize particles. *Soils and Foundations*, 31 (3), 161–166.

Hubei E'xi Institute of Geological Engineering Investigation (2002) *Report of the Field Investigation on the Tanping Landslide in the Three Gorgers Area*. Report number: 209.

Hungr, O. & Morgenstern, N.R. (1984) High velocity ring shear tests on sand. *Geotechnique*, 34 (3), 415–421.

Hutchinson, J.N. (1988) General report: Morphological and geological parameters of landslides in relation to geology and hydrogeology. In: Nonnare, C. (ed.) *Proc 5th Int Symp on Landslides, 10–15 Jul 1988, Lausanne, Switzerland*, Vol. 1. Rotterdam, Balkema. pp. 3–36.

Hutchinson, J.N., Somerville, S.H. & Petley, D.J. (1973) A landslide in periglacially disturbed Etruria Mari at Bury Hill. *Quarterly Journal of Engineering Geology and Hydrogeology*, 6, 377–404.

Hutchinson, J.N., Brohead, E.N. & Lupini, J.F. (1980) Additional observations on the Folkesone Warren landslides. *Quarterly Journal of Engineering Geology and Hydrogeology*, 13, 1–31.

Hvorslev, M.J. (1936) A ring shearing apparatus for the determination of the shearing resistance and plastic flow of soils. In: *Proc 1st Int Conf on Soil Mech Foud Eng, Boston, USA*, Vol. 2. pp. 125–129.

Hvorslev, M.J. (1939) Torsion shear tests and their place in the determination of the shearing resistance of soils. In: *Proc Am Soc Test Mater, Conshohocken, USA*, Vol. 39. pp. 999–1022.

Hvorslev, M.J. (1960) Physical components of the shear strength of saturated clays. In: *Proc Shear Strength of Cohesive Soils Conf, Jun 1960*. Colorado, ASCE. pp. 169–273.

Hvorslev, M.J. & Kaufman, R.I. (1952) *Torsion Shear Apparatus and Testing Procedures*. Vicksburgh, MS, US Army Waterways Experimental Station Corps of Engineers. Bulletin number: 38.

Hyslip, J.P. & Vallejo, L.E. (1997) Fractal analysis of roughness and size distribution of granular materials. *Engineering Geoloy*, 48, 231–244.

Idriss, I.M. (1985) Evaluating seismic risk in engineering practice. In: *Proc 11th Int Conf on Soil Mech Foud Eng, 12–16 Aug 1985, San Francisco, USA*, Vol. 1. Rotterdam, Balkema, A.A. pp. 255–320.

Igwe, O., Sassa, K. & Wang. F. (2007) The influence of grading on the shear strength of loose sands in stress-controlled ring shear tests. *Landslides*, 4, 43–51.

Irfan, T.Y. & Tang, K.Y. (1992) *An Engineering Geological Characterization of Colluvium in Hong Kong*. Hong Kong, Geotechnical Engineering Office. Technical note number: TN 4/92.

Irfan, T.Y. & Tang, K.Y. (1995) *Effect of the Coarse Fractions on the Shear Strength of Colluvium*. Hong Kong, Government Publications Center.

Jia, X. & Williams, R.A. (2001) A packing algorithm for particles of arbitrary shapes. *Powder Technology*, 120, 175–186.

Jovicic, V. & Coop, M.R. (1997) Stiffness of coarse-grained soils at small strains. *Geotechnique*, 47 (3), 545–561.

Kakou, B.G., Shimizu, H. & Nishimura, S. (2001) Residual strength of colluvium and stability analysis of farmland slope. *Agricultural Engineering International: CIGR Journal of Scientific Research and Development*, 3, 1–12.

Kalteziotis, N. (1993) The residual shear strength of some Hellenic clayey soils. *Geotechnical and Geological Engineering*, 11, 125–145.

Kanji, M.A. & Wolle, C.M. (1977) Residual strength – New testing and microstructure. In: *Proc 9th Int Conf of Soil Mech Foud Eng, Tokyo, Japan*, Vol. 1. pp. 153–154.

Kelly, R., Airey, D.W. & Tabucanon, J.T. (2003) Design and performance of a 1 m diameter ring shear apparatus. *Geotechnical Testing Journal*, 26 (4), 1–6.

Kenney, T.C. (1967) Slide behaviour and shear resistance of a quick clay determined from a study of the landslide at Selnes, Norway. In: *Proc Geotech Conf, Oslo, Norway*. Oslo, Norwegian Geotechnical Institute. pp. 57–64.

Kenney, T.C. (1976) The influence of mineral composition on the residual strength of natural soils. In: *Proc Geotech Conf, Oslo, Norway*, Vol. 1. Rotterdam, Balkema, A.A. pp. 123–129.

Kenny, T.C. (1977) Residual strengths of mineral mixtures. In: *Proc 9th Int Conf on Soil Mech Found Eng, Tokyo, Japan*. Tokyo, Japanese Society of Soil Mechanics and Foundation Engineering. pp. 155–160.

Kim, D.Y., Chun, B.S. & Yang, J.S. (2006) Development of a direct shear apparatus with rock joints and its verification tests. *Geotechnical Testing Journal*, 29 (5), 1–9.

La Gatta, D.P. (1970) *Residual Strength of Clays and Clay Shales by Rotation Shear Tests*. Cambridge, MA, Harvard. Soil Mechanics Series number: 86.

Lambe, T.W. (1985) Amuay landslide. In: *Proc 11th Int Conf of Soil Mech Found Eng, 12–18 Aug 1985, San Francisco, USA*. Rotterdam, Balkema.

Lemos, L.J.L. (1986) *The Effect of Rate of Shear on Residual Strength Soil*. PhD Thesis. London, University of London.

Lemos, L.J.L., Skempton, A.W. & Vaughan, P.R. (1985) Earthquake loading of shear surfaces in slopes. In: *Proc 11th Int Conf of Soil Mech Found Eng, 12–16 Aug 1985, San Francisco, USA*, Vol. 4. Rotterdam, Balkema A.A. pp. 1955–1962.

Leroueil, S. (2001) Natural slopes and cuts: Movement and failure mechanisms. *Geotechnique*, 51 (3), 197–243.

Li, J.J., Xie, S.Y. & Kuang, M.S. (2001) Geomorphic evolution of the Yangtze Gorges and the time of their formation. *Geomorphology*, 41, 125–135.

Li, L. (2002) Geological hazards and prevention in the Three Gorges dam area (in Chinese). *Landslide Resources*, 4, 4–7.

Li, T. & Wang, S. (1992) *Landslide Hazards and Their Mitigation in China* (in Chinese). Beijing, Science Press.

Li, Y.R. (2013) Effects of particle shape and size distribution on the shear strength behavior of composite soils. *Bulletin of Engineering Geology and the Environment*, 72 (3), 371–381.

Li, Y.R. & Aydin, A. (2010) Behavior of rounded granular materials in direct shear: Mechanics and quantification of fluctuations. *Engineering Geology*, 115 (1–2), 96–114.

Li, Y.R. & Aydin, A. (2013) Shear zone structures and stress fluctuations in large ring shear tests. *Engineering Geology*, 167, 6–13.

Li, Y.R., Aydin, A., Xu, Q. & Chen, J. (2012) Constitutive behavior of binary mixtures of kaolin and glass beads in direct shear. *KSCE Journal of Civil Engineering*, 16 (7), 1152–1159.

Li, Y.R., Wen, B., Aydin, A. & Ju, P. (2013a) Ring shear tests on slip zone soils of three giant landslides in the Three Gorges Project area. *Engineering Geology*, 154 (2), 106–115.

Li, Y.R., Huang, R.Q. & Chan, L.S. (2013b) Effects of particle shape on shear strength of clay–gravel mixture. *KSCE Journal of Civil Engineering*, 17 (4), 704–709.

Li, Y.R., Chan, L.S. & Yeung, A.T. (2013c) Effects of test conditions on shear behavior of composite soil. *Proceedings of the Institution of Civil Engineers – Geotechnical*, 166 (3), 310–320.

Liao, Q.L., Li, X., Lee, S.T. & Dong, Y.H. (2005) Occurrence geology and geomorphy characteristics and origin of Qianjiangping landslide in Three Gorges Reservoir area and study on ancient landslide criterion (in Chinese). *Chinese Journal of Rock Mechanics and Engineering*, 24 (17), 3146–3153.

Liu, J.G., Mason, P.J., Clerici, N., Chen, S., Davis, A., Miao, F., Deng, H. & Liang, L. (2004) Landslide hazard assessment in the Three Gorges area of the Yangtze River using ASTER imagery: Zigui–Badong. *Geomorphology*, 61, 171–187.

Liu, X.L., Loo, H., Min, H., Deng, J.H., Tham, L.G. & Lee, C.F. (2006) Shear strength of slip soils containing coarse particles of Xietan landslide. In: Han, J., Yin, J.H., White, W.D. & Lin, G.M. (eds.) *Proc Sess GeoSh, 6–8 Jun 2006, Shanghai, China*. pp. 195–207.

Lupini, J.F. (1980) *The Residual Strength of Soils*. PhD Thesis. London, University of London.

Lupini, J.F., Skinner, A.E. & Vaughan, P.R. (1981) The drained residual strength of cohesive soils. *Geotechnique*, 31 (2), 181–213.

Ma, Z.S., Zhang, Q., Zhu, R. & Jiang, Z.H. (2005) The basic characters of mountain disasters and relationship between landslide and rainfall in the area of Three Gorges Reservoir (in Chinese). *Journal of Mountain Science*, 23 (3), 319–326.

Martins, J.P. (1983) *Shaft Resistance of Axially Loaded Piles in Clay*. PhD Thesis. London, University of London.

Mason, R. (1999) The Three Gorges Dam of the Yangtze River, China: Engineering geology in China. *Geology Today*, 15 (1), 30–33.

McGeary, R.K. (1961) Mechanical packing of spherical particles. *Journal of the American Ceramic Society*, 44, 513–522.

Meloy, T.P. (1977) Fast Fourier Transforms applied to shape analysis of particle silhouettes to obtain morphological data. *Powder Technology*, 17, 27–35.

Mesri, G. & Abdel-Ghaffar, M.E.M. (1993) Cohesion intercept in effective stress–stability analysis. *Journal of Geotechnical Engineering*, 119 (8), 1229–1249.

Mesri, G. & Cepeda-Diaz, A.F. (1986) Residual shear strength of clays and shales. *Geotechnique*, 36, 269–274.

Mesri, G. & Shahien, M. (2003) Residual shear strength mobilized in first-time slope failures. *Journal of Geotechnical and Geoenvironmental Engineering*, 129 (1), 12–31.

Miller, E.A. & Sowers, G.F. (1957) The strength characteristics of soil-aggregate mixture. *Highway Research Board Bulletin*, 183, 16–23.

Mitchell, J.K. (1976) *Fundamentals of Soil Behaviour*. New York, NY, John Wiley & Sons.

Mitchell, J.K. (1993) *Fundamentals of Soil Behaviour*. 2nd edition. New York, NY, John Wiley & Sons.

Miura, K., Maeda, K., Furukawa, M. & Toki, S. (1998) Mechanical characteristics of sands with different primary properties. *Soils and Foundations*, 38, 159–172.

Morgenstern, N.R. & Tchalenko, J.S. (1976a) Microscopic structures in kaolin subjected to direct shear. *Geotechnique*, 17 (4), 309–329.

Morgenstern, N.R. & Tchalenko, J.S. (1976b) Microstructural observations on shear zones from slips in natural clays. In: *Proc Geotech Conf, Oslo, Norway*, Vol. 1. Oslo, Norwegian Geotechnical Institute. pp. 147–152.

Nakata, Y., Kato, Y., Hyodo, M., Hyde, A.F.L. & Murata, H. (2001) One-dimensional compression behaviour of uniformly graded sand related to single particle crushing strength. *Soils and Foundations*, 41 (2), 39–51.

Ng, C.W.W. & Chui, A.C.F. (2003) Laboratory study of loose saturated and unsaturated decomposed granitic soil. *Journal of Geotechnical and Geoenvironmental Engineering*, 129 (6), 550–559.

Oda, M. & Kazama, H. (1998) Microstructure of shear bands and its relation to the mechanism of dilatancy and failure of dense granular soils. *Geotechnique*, 48 (4), 465–481.

Oda, M. & Konishi, J. (1974) Microscopic deformation mechanism of granular material in simple shear. *Soils and Foundations*, 14 (4), 25–38.

Parathiras, A.N. (1994) *Displacement Rate Effects on the Residual Strength of Soils*. PhD Thesis. London, University of London.

Patwardhan, A.S., Rao, J.S. & Gaidhane, R.B. (1970) Interlocking effects and shearing resistance of boulders and large size particles in a matrix of fines on the basis of large scale direct shear tests. In: *Proc 2nd SE Asian Conf on Soil Mech, 11–15 Jun 1970, Singapore*. Pathumthani, Southeast Asian Geotechnical Society. pp. 265–273.

Peck, R.B. (1967) Stability of natural slopes. *Journal of the Soil Mechanics and Foundations Division – ASCE*, 93 (SM4), 403–417.

Petley, D.J. (1966) *The Shear Strength of Soils at Large Strains*. PhD Thesis. London, University of London.

Pitman, T.D., Robertson, P.K. & Sego, D.C. (1994) Influence of fines on the collapse of loose sands. *Canadian Geotechnical Journal*, 31, 728–739.

Poulos, S.J. (1981) The steady state of deformation. *Journal of Geotechnical and Geoenvironmental Engineering*, 107 (5), 553–562.

Prakasha, K.S. & Chandrasekaran, V.S. (2005) Behavior of marine sand–clay mixtures under static and cyclic triaxial shear. *Journal of Geotechnical and Geoenvironmental Engineering*, 131 (2), 213–222.

Ramiah, B.K., Dayalu, N.K. & Purushothamaraj, P. (1970) Influence of chemicals on residual strength of silty clay. *Soils and Foundations*, 10 (1), 25–36.

Rogers, C.D.F., Dijkstra, T.A. & Smalley, I.J. (1994) Particle packing from an earth science point of view. *Earth Science Review*, 36, 59–82.

Roscoe, K.H. & Burland, J.B. (1968) On the generalized stress–strain behaviour of "wet" clay. In: Heyman, J. & Leckie, F.A. (eds.) *Engineering Plasticity*. Cambridge, Cambridge University Press. pp. 535–609.

Roscoe, K.H., Schofield, A.W. & Wroth, C.P. (1958) On the yielding of soils. *Geotechnique*, 8 (1), 22–53.

Roscoe, K.H., Schofield, A.N. & Thurairajah, A. (1963) Yielding of clays in states wetter than critical. *Geotechnique*, 13 (3), 211–240.

Rothenburg, L. & Bathurst, R.J. (1989) Analytical study of induced anisotropy in idealized granular material. *Geotechnique*, 49, 601–614.

Santamrina, J.C. & Cho, G.C. (2004) Soil behavior: The role of particle shape. In: Jardine, R.J., Potts, D.M. & Higgins, K.G. (eds.) *Adv in Geotech Eng: The Skempton Conf, 29–31 Mar 2004, London, UK*. London, Thomas Telford. pp. 604–617.

Sassa, K. (1985) The mechanism of debris flow. In: *Proc 11th Int Conf of Soil Mech Found Eng, 12–16 Aug 1985, San Francisco, USA*. Rotterdam, Balkema, A.A. pp. 1173–1176.

Sassa, K. (1988) Geotechnical model for the motion of landslides. In: *Spec Lect 5th Int Sym on Landslides, 10–15 Jul 1988, Lausanne, Switzerland*, Vol. 1. Rotterdam, Balkema, A.A. pp. 37–55.

Sassa, K. (1992) Access to the dynamics of landslides during earthquakes by a new cyclic loading ring-shear apparatus. Keynote paper. In: Bell, D.H. & Balkema, A.A. (eds.) *Proc 6th Int Symp on Landslides, 10–14 Feb 1992, Christchurch, New Zealand*, Vol. 3. Rotterdam, Balkema, A.A. pp. 1919–1939.

Sassa, K. (1996) Prediction of earthquake induced landslides. In: *Spec Lect 7th Int Symp on Landslides, 17–21 Jun 1996, Trondheim, Norway*. Rotterdam, Balkema, A.A. pp. 115–132.

Sassa, K., Wang, G., Fukuoka, H., Wang, F., Occhai, T., Sugiyama, M. & Sekiguchi, T. (2004) Landslide risk evaluation and hazard zoning for rapid and long-travel landslides in urban development areas. *Landslides*, 1 (3), 221–235.

Schofield, A. (2005) *Disturbed Soil Properties and Geotechnical Design*. London, Thomas Telford.

Schofield, A. & Wroth, P. (1968) *Critical State Soil Mechanics*. London, McGraw-Hill.

Schuster, R.L. (1996) Socioeconomic significance of landslides. In: Turner, A.K. & Schuster, R.L. (eds.) *Landslide, Investigation and Mitigation*. Transport Research Board Manual. pp. 12–35.

Seed, H.B., Woodward, R.J. & Lundgren, R. (1964) Clay mineralogical aspects of the Atterberg limits. *Journal of ASCE*, 90 (SM4), 107–131.

Seycek, J. (1978) Residual shear strength of soils. *Bulletin of Engineering Geology and the Environment*, 17 (1), 73–75.

Shakoor, A. & Cook, B.D. (1990) The effect of stone content, size and shape on the engineering properties of a compacted silty clay. *Bulletin of the Association of Engineering Geologists*, 27 (2), 245–253.

Sharma, B. & Bore, P.K. (2003) Plastic limit, liquid limit and undrained shear strength of soil-reappraisal. *Journal of Geotechnical and Geoenvironmental Engineering*, 129 (8), 774–777.

Shelly, T.L. & Daniel, D.E. (1993) Effect of gravel on hydraulic conductivity of compacted soil liners. *Journal of Geotechnical Engineering – ASCE*, 119, 54–68.

Shimizu, M. (1997) Strain fields in direct shear box tests on a metal-rods model of granular soils. In: Asaoka, A., Adachi, T. & Oka, F. (eds.) *Deformation and Progressive Failure in Geomechanics*. Pergamon, Elsevier Science. pp. 151–156.

Shimobe, S. & Moroto, N. (1995) A new classification chart for sand liquefaction. In: Ishihara, K. (ed.) *Proc 1st Int Conf on Earthquake Geotech Eng, 14–16 Nov 1995, Tokyo, Japan*. Rotterdam, Balkema, A.A. pp. 315–320.

Simoni, A. & Houlsby, G.T. (2006) The direct shear strength and dilatancy of sand–gravel mixtures. *Geotechnical and Geological Engineering*, 24, 523–549.

Skempton, A.W. (1956) *Alexandre Collin (1808–1890) and His Pioneer Work in Soil Mechanics*. Preface to the translation by W.R. Schriever of Alexandre Collin (1846) book. Toronto, ON, University of Toronto Press.

Skempton, A.W. (1964) The long term stability of clay slopes. *Geotechnique*, 14 (2), 77–101.

Skempton, A.W. (1985) Residual strength of clays in landslides, folded strata and the laboratory. *Geotechnique*, 35 (1), 3–8.

Skempton, A.W. & Delory, F.A. (1957) Stability of natural slopes in London clay. In: *Proc 4th Int Conf on Soil Mech Found Eng, 12–24 Aug 1957, London, UK*, Vol. 2. pp. 378–381.

Skempton, A.W. & Petley, D.J. (1967) The strength along structural discontinuities in stiff clays. In: *Proc Geotech Conf, Oslo, Norway*, Vol. 2. Oslo, Norwegian Geotechnical Institute. pp. 29–46.

Skinner, A.E. (1969) *The Effect of High Pore Water Pressures on the Mechanical Behaviour of Sediments*. PhD Thesis. London, University of London.

Smart, P. (1970) Residual shear strength. *Journal of ASCE*, 96 (SM6), 2181–2183.

Smith, S.W. (1997) *The Scientist and Engineer's Guide to Digital Signal Processing*. California, California Technical Publishing.

Spear, D.A. & Taylor, R.K. (1972) The influence of weathering on the composition and engineering properties of in situ Coal Measures rocks. *International Journal of Rock Mechanics and Mining*, 9, 729–756.

Stark, T.D., Choi, H. & McCone, S. (2005) Drained shear strength parameters for analysis of landslides. *Journal of Geotechnical and Geoenvironmental Engineering*, 131, 575–588.

Stark, T.D. & Eid, H.T. (1994) Drained residual strength of cohesive soils. *Journal of Geotechnical and Geoenvironmental Engineering*, 120 (5), 856–871.

Suzuki, M., Yamamoto, T., Sasanishi, T. & Sugawara, M. (2000) Residual strength characteristics of pure clay minerals prepared with different salinity. *Memoirs of the Faculty of Engineering, Yamaguchi University*, 54 (1), 11–15.

Suzuki, M., Yamamoto, T. & Tanikawa, K. (2001) Variation in residual strength of clay with sharing speed. *Memoirs of the Faculty of Engineering, Yamaguchi University*, 52 (1), 45–49.

Suzuki, M., Tsuzuki, S. & Yamamoto, T. (2005) Physical and chemical index properties of residual strength of various soils. *Memoirs of the Faculty of Engineering, Yamaguchi University*, 56 (1), 1–10.

Taylor, D.W. (1948) Fundamentals of soil mechanics. *Soil Science*, 66 (2), 21–133.

Terzaghi, K. (1925) *Erdbaumechanik auf bodenphysikalischer Grundlage*. Franz Deuticke, Leipzig und Wein.

Thevanayagam, S. & Mohan, S. (2000) Intergranular state variables and stress–strain behaviour of silty sands. *Geotechnique*, 50 (1), 1–23.

Thornton, C. (2000) Numerical simulations of deviatoric shear deformation of granular media. *Geotechnique*, 50, 43–53.

Tika, T.E. & Hutchinson, J.N. (1999) Ring shear tests on soil from the Vaiont landslide surface. *Geotechnique*, 49 (1), 59–74.

Tika, T.E., Vaughan, P.R. & Lemos, L.J. (1996) Fast shearing of pre-existing shear zones in soil. *Geotechnique*, 46 (2), 197–233.

Tika, T.M. (1989) *The Effect of Rate of Shear on the Residual Strength of Soil*. PhD Thesis. London, University of London.

Ting, W.H., Mum, K.P. & Toh, C.T. (1982) Characteristics of a composite residual granite soil. In: Mcfeat-Smith, I. & Lumb, P. (eds.) *Proc 7th SE Asian Geotech Conf, 22–26 Nov 1982, Hong Kong, China*, Vol. 1. pp. 879–887.

Tiwari, B. & Marui, H. (2004) Objective-oriented multistage ring shear test for the shear strength of landslide soil. *Journal of Geotechnical and Geoenvironmental Engineering*, 130 (2), 217–222.

Tiwari, B. & Marui, H. (2005) A new method for the correlation of residual shear strength of the soil with mineralogy composition. *Journal of Geotechnical and Geoenvironmental Engineering*, 131 (9), 1139–1150.

Tiwari, B., Brandon, T.L., Marui, H. & Tuladhar, G.R. (2005) Comparison of residual shear strengths from back analysis and ring shear tests on undisturbed and remolded specimens. *Journal of Geotechnical and Geoenvironmental Engineering*, 131 (9), 1071–1079.

Townsend, F.C. & Gilbert, P.A. (1973) Tests to measure residual strengths of some clay shales. *Geotechnique*, 23 (2), 267–271.

Townsend, F.C. & Gilbert, P.A. (1974) *Engineering Properties of Clay Shales, Report 2: Residual Shear Strength and Classification Indexes of Clay Shales*. Vicksburg, Soils and Pavements Laboratory, US Army Engineering Waterways Experiment Station.

Townsend, F.C. & Gilbert, P.A. (1976) Effects of specimen type on the residual strength of clays and clay shales. In: *Soil Specimen Preparation for Laboratory Testing*. ASTM Special Technical Publication 599, American Society for Testing and Materials. pp. 43–65.

Vallejo, L.E. (1989) An extension of the particulate model of stability analysis for mudflows. *Soils and Foundations*, 29 (3), 1–13.

Vallejo, L.E. (2001) Interpretation of the limits in shear strength in binary granular mixtures. *Canadian Geotechnical Journal*, 38, 1097–1104.

Vallejo, L.E. & Mawby, R. (2000) Porosity influence on the shear strength of granular material–clay mixtures. *Engineering Geology*, 58, 125–136.

Vallejo, L.E. & Zhou, Y. (1994) The mechanical properties of simulated soil–rock mixtures. In: *Proc 13th Int Conf on Soil Mech Found Eng, 5–10 Jan 1994, New Delhi, India*. Rotterdam, Balkema, A.A. pp. 365–368.

Van der Merwe, D.H. (1964) The prediction of heave from the plasticity index and percentage of clay fraction of soils. *Civil Engineering*, 6, 103–107.

Varnes, D.J. (1978) Slope movement types and processes. In: Schuster, R.L. & Krizek, R.J. (eds.) *Landslides: Analysis and Control. Transportation Research Board Sp. Rep. No. 176*. New York, NY, National Academy of Sciences. pp. 11–33.

Vaughan, P.R. & Walbancke, H.J. (1975) The stability of cut and fill slopes in boulder clay. In: *Proc Symp Eng Behav Glacial Mate, Midland Soc Soil Mech, 21–23 Apr 1975, Birmingham, UK*. pp. 209–219.

Vaughan, P.R., Davachi, M.M., El Ghamrawy, M.K., Hamza, M.M. & Hight, D.W. (1976) Stability analysis of large gravity structures. In: *Proc 1st Int Conf on the Behav Off-Shore Struct, 2–5 Aug 1976, Trondheim, Norway*, Vol. 1. Trondheim, Norwegian Institute of Technology. pp. 467–480.

Vaughan, P.R., Hight, D.W., Sodha, V.G. & Walbancke, H.J. (1978) *Factors Controlling the Stability of Clay Fills in Britain. Clay Fills*. London, Institution of Civil Engineers.

Voight, B. (1973) Correlation between Atterberg plasticity limits and residual shear strength of natural soils. *Geotechnique*, 23 (2), 265–267.

Wafid, M., Sassa, K., Fukuoka, H. & Wang, G. (2004) Evolution of shear-zone structure in undrained ring shear tests. *Landslides*, 1 (2), 101–112.

Wan, Y.S. & Kwong, J. (2002) Shear strength of soils containing amorphous clay-size materials in a slow-moving landslide. *Engineering Geology*, 65, 293–303.

Wang, F.W., Zhang, Y.M., Huo, Z.T., Matsumoto, T. & Huang, B.L. (2004) The July 14, 2003 Qianjiangping landslide, Three Gorges Reservoir, China. *Landslides*, 1, 157–162.

Wang, S.Q. (1992) Modeling landslide distribution on loess soils in China: An investigation. *International Journal of Geographical Information Systems*, 6 (5), 391–405.

Wen, B.P. & Aydin, A. (2003) Microstructural study of a natural slip zone: Quantification and deformation history. *Engineering Geology*, 68 (3–4), 289–317.

Wen, B.P. & Aydin A. (2005) Mechanism of a rainfall-induced slide-debris flow: Constraints from microstructure of its slip zone. *Engineering Geology*, 78, 69–88.

Wen, B.P., Aydin, A., Duzgoren-Aydin, N.S., Li, Y.R., Chen, H.Y. & Xiao, S.D. (2007) Residual strength of slip zones of large landslides in the Three Gorges Area, China. *Engineering Geology*, 93, 82–98.

Wesley, L.D. (1977) Shear strength properties of halloysite and allophone clays in Java, Indonesia. *Geotechnique*, 27 (2), 125–136.

Wesley, L.D. (2003) Residual strength of clays and correlations using Atterberg limits. *Geotechnique*, 53 (7), 669–672.

Williams, A.A.B. & Donaldson, G. (1980) Building on expansive soils in South Africa. In: Snethen, D. (ed.) *Proc 4th Int Conf on Expansive Soils, 16–18 Jun 1980, Denver, USA*, Vol. 2. New York, NY, ASCE. pp. 834–848.

Wu, S.R., Shi, L., Wang, R.J, Tan, C.X., Hu, D.G., Mei, Y.T. & Xu, R.C. (2001) Zonation of the landslide hazards in the forereservoir region of the Three Gorges Project on the Yangtze River. *Engineering Geology*, 59, 51–58.

Xie, S.Y., Zhang, N.X. & Xu, B. (1995) Probability analysis of precipitation-induced slide of typical landslides in Three Gorges Area of The Yangtze River. *Journal of Engineering Geology*, 3 (2), 60–69.

Xu, Q., Huang, R.Q., Cheng, Q.G., Ding, X.M., Li, D.S. & Li, Y.R. (2003) Study on soil properties in sliding zone of Xietan landslide on the Three Gorges reservoir (in Chinese). *Journal of Engineering Geology*, 11 (4), 354–359.

Xu, Z., Zhou, G., Liu, Z., Zhou, J. & Tian, Q. (2007) Correcting method and error analysis for sample area in direct shear test. *Journal of China University of Mining Technology*, 36 (5), 658–662.

Yagi, N., Yatabe, R. & Enoki, M. (1991) Stability analyses for landslides using ring shear results. In: Bell, D.H. (ed.) *Proc 6th Int. Symp. on Landslides, Christchurch, New Zealand*, Vol. 1. Rotterdam, Balkema, A.A. pp. 607–612.

Yan, R.M.C. (2005) Measurement of particle shape using digital imaging techniques. *Journal of Physics: Conference Series*, 15, 177–182.

Yoshida, N. & Hosokawa, K. (2004) Compression and shear behavior of mudstone aggregates. *Journal of Geotechnical and Geoenvironmental Engineering*, 130 (5), 519–524.

Zaruba, Q. & Mencl, V. (1982) *Landslides and Their Control*. 2nd edition. New York, NY, Elsevier Scientific.

Zhao, C. (1996) River capture and original of the Yangtze River (in Chinese). *Journal of Changchun University of Earth Science*, 26, 428–433.

# Subject index